# Lecture Notes in Mathematics    1508

Editors:
A. Dold, Heidelberg
B. Eckmann, Zürich
F. Takens, Groningen

Matti Vuorinen (Ed.)

# Quasiconformal Space Mappings

A collection of surveys 1960-1990

Springer-Verlag

Berlin Heidelberg New York
London Paris Tokyo
Hong Kong Barcelona
Budapest

Editor

Matti Vuorinen
Mathematics Department
University of Helsinki
Hallitusk. 15
SF-00100 Helsinki, Finland

Mathematics Subject Classification (1991): 30C65, 30F35, 31B15, 33E05, 35J60, 46E30, 46E35, 49-02, 49N60, 54C10, 54E35, 57M10, 58C15, 73C50

ISBN 3-540-55418-1 Springer-Verlag Berlin Heidelberg New York
ISBN 0-387-55418-1 Springer-Verlag New York Berlin Heidelberg

Typesetting: Camera ready by author/editor
Printing and binding: Druckhaus Beltz, Hemsbach/Bergstr.
46/3140-543210 - Printed on acid-free paper

Quasiconformal Space Mappings

A collection of surveys 1960 – 90

## CONTENTS

# Preface

Quasiconformal (qc) maps of the complex plane were introduced by H. Grötzsch in 1928 as a generalization of the notion of conformal maps. Beginning in the mid thirties, far-reaching progress in the theory was made by the work of O. Teichmüller and L. V. Ahlfors. In three-dimensional euclidean space these maps occurred in the work of M. A. Lavrent'ev [Lav] in 1938, but the modern era of the theory of qc homeomorphisms in euclidean $n$–space began around 1960, when several papers of E. D. Callender [CA], F. W. Gehring [G1], C. Loewner [L], Yu. G. Reshetnyak [R1], B. V. Shabat [Sh], and J. Väisälä [V1] appeared almost simultaneously. An important breakthrough in the subject was the generalization to quasiregular (qr) mappings, not necessarily injective, whose basic theory was pioneered by Reshetnyak [R1], [R2], [R3]. During the past three decades a rich theory of these maps has developed from the work of many mathematicians. The results they have obtained show that qc and qr maps are natural generalizations of conformal mappings and analytic functions of one complex variable, respectively. Various aspects of the theory are covered by the monographs [C], [V2], [R2], [R3], [Ri2], [Vu] as well as by the surveys [BM], [G2], [I1], [LF], [Ri1], [V3].

Qc maps interpolate between bilipschitz maps and general homeomorphisms. Two-dimensional qc maps occur in D. Sullivan's work [S] on dynamical systems, and qc maps are also used in the study of fractals. The remarkable recent results of Donaldson and Sullivan [DS] on 4–manifolds and the work of T. Iwaniec [I2] based on their paper give promise of significant progress on qc and qr maps in the near future. Other recent results appear in the work of Friedman and He [FH].

The present volume is the first collection of surveys devoted to the theory of quasiconformal space mappings. Individual papers are surveys of their respective topics. The paper of Gehring is a reprint of his Plenary Address [G3] delivered in the ICM86 in Berkeley and is reprinted here with the kind permission of the American Mathematical Society. The paper of Reshetnyak provides a new proof of a well-known result; thus his paper is partly expository, but it also contains original results. The papers published here cover a wide spectrum of results concerning qc maps, but some topics such as the connection of qc maps with geometric topology [TV] are not covered at all. Some topics of geometric function theory related to qc and qr maps are surveyed in [D].

This volume would not have been possible without the cooperation of the authors, to whom I express my sincere appreciation. Likewise, I want to thank the referees for their valuable contributions to the improvement of the papers. I am very much indebted to G. D. Anderson of Michigan State University for his generous help in the preparation of this volume.

December 1991

Matti Vuorinen

## References

[BM]   A. BAERNSTEIN and J. J. MANFREDI: *Topics in Quasiconformal Mapping.* – Topics in modern harmonic analysis. Proc. Seminar held in Torino and Milano, May–June 1982, Vol. II, 849– 862.

[Ca]   E. D. CALLENDER: *Hölder continuity of $n$ – dimensional quasiconformal mappings.* – Pacific J. Math. 19 (1960), 499–515.

[C]    P. CARAMAN: *$n$ –Dimensional Quasiconformal (QCf) Mappings.* – Editura Academiei Române, Bucharest, Abacus Press, Tunbridge Wells Haessner Publishing, Inc., Newfoundland, New Jersey, 1974.

[DS]   S. K. DONALDSON and D. P. SULLIVAN: *Quasiconformal 4-manifolds.* – Acta Math. 163 (1990), 181–252.

[D]    D. DRASIN, C. EARLE, F. W. GEHRING, I. KRA, and A. MARDEN (EDITORS): *Holomorphic Functions and Moduli I, II.* – Springer–Verlag, · 1988.

[FH]   M. H. FRIEDMAN and Z.-X. HE: *Divergence–free fields: Energy and asymptotic crossing number.* – Ann. Math. 134 (1991), 189–229.

[G1]   F. W. GEHRING: *Symmetrization of rings in space.* – Trans. Amer. Math. Soc. 101 (1961), 499–519.

[G2]   F. W. GEHRING: *Quasiconformal mappings in space.* – Bull. Amer. Math. Soc. 69 (1963), 146–164.

[G3]   F. W. GEHRING: *Topics in quasiconformal mappings.* – Proc. Internat. Congr. Math. (Berkeley, California 1986), Vol. 1, 62–80, AMS, 1987.

[I1]   T. IWANIEC: *Some aspects of partial differential equations and quasiregular mappings.* – Proc. Internat. Congr. Math. (Warsaw, 1983), Vol. 2, 1193–1208, PWN, Warsaw, 1984.

[I2]   T. IWANIEC: *$p$– harmonic tensors and quasiregular mappings.* – Ann. Math. (to appear).

[Lav]  M. A. LAVRENT'EV: *Sur un critère différentiel des transformations homéomorphes des domaines à trois dimensions.* – Dokl. Akad. Nauk SSSR 22 (1938), 241–242.

[LF]   J. LELONG–FERRAND: *Problèmes de géométrie conforme.* – Proc. Internat. Congr. Math. (Vancouver, B. C., 1974), Vol. 2, 13–19, Canad. Math. Congress, Montreal, Que., 1975.

[L]    C. LOEWNER: *On the conformal capacity in space.* – J. Math. Mech. 8

(1959), 411–414.

[R1]   YU. G. RESHETNYAK: *A sufficient condition for Hölder continuity of a mapping.* – Soviet Math. Doklady 1 (1960), 76–78.

[R2]   YU. G. RESHETNYAK: *Space Mappings with Bounded Distortion.* – Translations of Mathematical Monographs Vol. 73, Amer. Math. Soc. Providence, Rhode Island, 1989, ( A review: Bull. Amer. Math. Soc. 24 (1991), 408–415).

[R3]   YU. G. RESHETNYAK: *Stability theorems in geometry and analysis.* (Russian). – Izdat. "Nauka", Sibirsk. Otdelenie, Novosibirsk, 1982.

[Ri1]  S. RICKMAN: *Recent advances in the theory of quasiregular mappings.* – Proc. 19th Nordic Congress of Math., Reykjavik 1984, 116–125, 1985.

[Ri2]  S. RICKMAN: *Quasiregular Mappings.* – (Monograph to appear).

[Sh]   B. V. SHABAT: *On the theory of quasiconformal mappings in space.* – Soviet Math. 1 (1960), 730–733.

[S]    D. SULLIVAN: *Quasiconformal homeomorphisms in dynamics, topology, and geometry.* – Proc. Internat. Congr. Math. (Berkeley, CA, 1986), Vol. 2, 1216–1228, AMS, 1987.

[TV]   P. TUKIA and J. VÄISÄLÄ: *Quasiconformal extension from dimension n to n+1.* – Ann. of Math. 115 (1982), 331– 348.

[V1]   J. VÄISÄLÄ: *On quasiconformal mappings in space.* – Ann. Acad. Sci. Fenn. Ser. A I 298 (1961), 1–36.

[V2]   J. VÄISÄLÄ: *Lectures on n –Dimensional Quasiconformal mappings.* – Lecture Notes in Math. Vol. 229, Springer-Verlag, Berlin– Heidelberg– New York, 1971.

[V3]   J. VÄISÄLÄ: *A survey of quasiregular maps in $R^n$.* – Proc. Internat. Congr. Math. (Helsinki, 1978), Vol. 2, 685–691, Acad. Sci. Fennica, Helsinki, 1980.

[Vu]   M. VUORINEN: *Conformal Geometry and Quasiregular Mappings.* – Lecture Notes in Math. Vol. 1319, Springer-Verlag, Berlin–Heidelberg–New York, 1988, (A review: Bull. Amer. Math. Soc. 21 (1989), 354–360).

Quasiconformal Space Mappings
– A collection of surveys 1960–1990
Springer-Verlag (1992), 1–19
Lecture Notes in Mathematics Vol. 1508

# CONFORMAL INVARIANTS, QUASICONFORMAL MAPS, AND SPECIAL FUNCTIONS

G. D. Anderson, M. K. Vamanamurthy, and M. Vuorinen

Michigan State University, East Lansing, MI 48824, USA,

University of Auckland, Auckland, New Zealand,

University of Helsinki, SF-00100 Helsinki, Finland.

## 1. Introduction

The study of geometry by means of invariants was envisaged by B. Riemann, F. Klein, and H. Poincaré in the nineteenth century. In the framework of geometric function theory these ideas were soon found to be fruitful, and particularly invariant metrics became a standard tool wielded by H. Schwarz, C. Carathéodory, and others. In 1928 H. Grötzsch [Gr] introduced a new invariant, the modulus of a quadrilateral, which enabled him to study a new class of mappings, now known as quasiconformal mappings. Ten years later O. Teichmüller [T] applied the results and methods of Grötzsch to create the foundations of what is now known as the theory of Teichmüller spaces. In 1950 there appeared a celebrated paper of L. V. Ahlfors and A. Beurling [AhB], in which they introduced a new conformal invariant called the extremal length of a curve family. The reciprocal of the extremal length is called the modulus. Their seminal work opened new perspectives and attracted the attention of many mathematicians. J. Hersch [H] (see also [Ke]) obtained connections between extremal length and harmonic measure, B. Fuglede [F] extended the notion of the extremal length to euclidean $n$–space, and J. Jenkins [J] applied these methods to the study of univalent functions. Although higher-dimensional quasiconformal mappings had first been considered in 1938–1941 by M. A. Lavrentiev [La], A. Markushevich [Ma], and M. Kreines [Kr], an extensive study of these mappings began only in 1959 onwards by several authors, especially C. Loewner [Lo], F. W. Gehring [G1, G2], J. Väisälä [V1] (see also [GV]), Yu. G. Reshetnyak [Re1, Re2], and B. V. Shabat [Sh]. The extremal length method became a widely used and crucial tool in the study of geometric properties of quasiconformal and quasiregular mappings both in the plane and in space [LV, V2, R, Vu2].

A homeomorphism $f : G \to G'$, where $G$ and $G'$ are domains in $\overline{\mathbf{R}}^n = \mathbf{R}^n \cup \{\infty\}$, is called $K$–quasiconformal, $K \geq 1$, if for every curve family $\Gamma$ in $G$ [V1]

$$(1.1) \qquad\qquad M(\Gamma)/K \leq M(f(\Gamma)) \leq KM(\Gamma),$$

where $M(\Gamma)$ denotes the modulus of $\Gamma$ (cf. §2 below). For $K = 1$ one obtains the class of conformal mappings, under which the modulus is invariant. The basic difference between the dimensions $n = 2$ and $n \geq 3$ is that by Riemann's mapping theorem plane conformal mappings constitute a very wide and flexible class, while by Liouville's theorem conformal mappings in space $\mathbf{R}^n$, $n \geq 3$, coincide with Möbius transformations and are thus very rigid. The basic results of the geometric theory of quasiconformal mappings are given in the well-known books of P. Caraman [Car] and J. Väisälä [V2]. There exists a rich theory of $K$–quasiconformal maps in $\mathbf{R}^n$, but few of these results are asymptotically sharp as $K \to 1$ for $n \geq 3$. In fact, a fundamental problem in the theory is to prove results for $K$–quasiconformal mappings in $\mathbf{R}^n, n \geq 3$, which are significant when applied to Möbius transformations. We shall briefly discuss below this problem from the point of view of distortion theory.

In spite of the simplicity of (1.1) it is not straightforward to take advantage of this condition in the study of geometric properties of $K$–quasiconformal mappings. For instance, no proof of Liouville's theorem based on (1.1) is known. The general difficulty of extracting geometric information about the action of quasiconformal mappings from (1.1) is connected with the fact that the definition of the modulus $M(\Gamma)$ is rather implicit and that $M(\Gamma)$ can be computed only for the simplest of curve families. For this reason one usually applies this modulus inequality only in cases where the curve families have relatively simple structure. Because of the Möbius invariance of this two-sided inequality it is natural to use the Möbius–invariant (absolute) cross ratio when studying the geometry of quasiconformal mappings.

This note is a survey of our recent results [AVV1–AVV8, Vu1–Vu4] on conformal invariants and special functions related to the moduli of curve families and quasiconformal mappings. Sharp estimates for such special functions play a major role in the study of quasiconformal mappings with maximal dilatation $K$ close to unity. The ultimate goal of the distortion theory is to prove results which in passage to the limit $K \to 1, n \geq 3$, would yield the Liouville theorem as a particular case. Because this goal is still largely beyond reach there is ample room for new research in this topic. For instance, it is a challenging task to find a quantitative form with concrete constants for the stability theory of Yu. G. Reshetnyak [Re3, p.252].

*Acknowledgement.* The research of the authors was supported in part by a grant from the Academy of Finland.

## 2. Capacity and modulus

We shall follow the notation in [V2], to which the reader is referred for the basic facts concerning quasiconformal mappings. Let $\Gamma$ be a family of curves in $\overline{\mathbf{R}}^n$. By

$\mathcal{F}(\Gamma)$ we shall mean the family of *admissible* functions, i.e. Borel-measurable functions $\rho : \mathbf{R}^n \to [0, \infty]$ such that

$$\int_\gamma \rho \, ds \geq 1$$

for each locally rectifiable curve $\gamma$ in $\Gamma$. Then the *modulus* of $\Gamma$ is defined by

$$(2.1) \qquad M(\Gamma) = \inf_{\rho \in \mathcal{F}(\Gamma)} \int_{\mathbf{R}^n} \rho^n \, dm,$$

where $m$ denotes $n$–dimensional Lebesgue measure. If $E, F, G$ are subsets of $\overline{\mathbf{R}}^n$, by $\Delta(E, F; G)$ we shall mean the family of all curves joining $E$ and $F$ in $G$.

Let $G$ be a proper subdomain of $\mathbf{R}^n$. Then we define two conformal invariants

$$(2.2) \qquad \mu_G(x, y) = \inf_{C_{xy}} M(\Delta(C_{xy}, \partial G; G))$$

and

$$(2.3) \qquad \lambda_G(x, y) = \inf_{C_x, C_y} M(\Delta(C_x, C_y; G)), \quad x \neq y,$$

for $x, y \in G$. In (2.2) the infimum is taken over all curves $C_{xy}$ joining $x$ and $y$ in $G$, while in (2.3) $C_z$ denotes a curve joining $z$ and $\partial G$ in $G$ [LF; Vu1, p.75; Vu2].

2.4. **Remarks.** (1) The functions $\lambda_G(x, y)$ and $\mu_G(x, y)$ are monotonic in the sense that $G_1 \subset G_2 \Rightarrow \lambda_{G_1}(x, y) \leq \lambda_{G_2}(x, y)$ and $\mu_{G_1}(x, y) \geq \mu_{G_2}(x, y)$ for all $x, y \in G_1$.
(2) $\mu_G$ and $\lambda_G^{-1/n}$ are metrics on $G$ [LF].

A *ring* $R$ is a domain in $\overline{\mathbf{R}}^n$ whose complement consists of two components. The *capacity* cap $R$ of a ring $R$ is defined as $M(\Gamma)$, where $\Gamma$ is the family of curves joining the components of the complement of $R$. The *modulus* of $R$ is defined in terms of the capacity as

$$\operatorname{mod} R = \left( \frac{\omega_{n-1}}{\operatorname{cap} R} \right)^{1/(n-1)},$$

where $\omega_{n-1}$ denotes the $(n-1)$–dimensional measure of the unit sphere in $\mathbf{R}^n$. A ring is said to be *extremal* if it has the largest modulus in some geometric class to which it belongs. Two particularly important extremal rings are the Grötzsch ring and the Teichmüller ring. For $x, y \in \mathbf{R}^n$ let $[x, y] = \{(1-s)x + sy : 0 \leq s \leq 1\}$ and for $x \in \mathbf{R}^n \setminus \{0\}$ let $[x, \infty] = \{sx : s \geq 1\} \cup \{\infty\}$. We let $e_i, i = 1, ..., n$, denote the coordinate unit vectors in $\mathbf{R}^n$.

The complementary components of the *Grötzsch ring* $R_{G,n}(s)$ are $\overline{B}^n$ and $[se_1, \infty]$, $s > 1$, while those of the *Teichmüller ring* $R_{T,n}(t)$ are $[-e_1, 0]$ and $[te_1, \infty]$, $t > 0$. We let

$$(2.5) \qquad \gamma_n(s) = \operatorname{cap} R_{G,n}(s), \quad \tau_n(t) = \operatorname{cap} R_{T,n}(t),$$

and

(2.6)
$$M_n(r) = \text{mod } R_{G,n}(s),$$

where $r = 1/s \in (0,1)$. For $n = 2$, $M_2(r)$ is also denoted by $\mu(r)$ (see (3.11)). The Grötzsch ring modulus has the following monotoneity property [AVV8]:

2.7. **Theorem.** *For each $n \geq 2$ the function*

$$M_n(r) + \log \frac{r}{1+r'}, \quad r' = \sqrt{1-r^2},$$

*is decreasing on $(0,1)$.*

From Theorem 2.7 it follows easily that $M_n(r) + \log r$ is strictly decreasing, as is well known [T, p. 627; G1; G2], and hence the limit

(2.8)
$$\log \lambda_n = \lim_{r \to 0+} (M_n(r) + \log r)$$

exists for each $n \geq 2$. The exact value of the Grötzsch ring constant $\lambda_n$ is known only when $n = 2$, in which case $\lambda_2 = 4$. For $n \geq 3$ the estimates $2e^{0.76(n-1)} < \lambda_n < 2e^{n-1}$ are known [AF]; in particular, $\lambda_n$ tends to $\infty$ with $n$. We also have the following estimates for $\lambda_n^{1-\alpha}$, where $\alpha = K^{1/(1-n)}$ [AVV5, Theorem 3.21]:

2.9. **Theorem.** *Let $n \geq 3$, $K > 1$, $\alpha = K^{1/(1-n)}$, and let $\lambda_n$ be the Grötzsch ring constant defined in (2.8). Then*

(2.10)
$$K^{0.76/\sqrt{K}} < \left(\frac{\lambda_n}{2}\right)^{1-\alpha} < K$$

and

(2.11)
$$\lim_{n \to \infty} \lambda_n^{1-\alpha} = K.$$

As a consequence of Theorem 2.7 we have

2.12. **Corollary.** *For each $n \geq 2$, $0 < r < 1$, $r' = \sqrt{1-r^2}$,*

$$\log \frac{1+r'}{r} < M_n(r) < \log \frac{\lambda_n(1+r')}{2r}.$$

By extremal length methods one may derive the following useful monotoneity properties [A; AVV5, Lemma 2.6(5)] and estimates [AVV5, (1.10)] for the Grötzsch ring modulus:

2.13. **Theorem.** *For $n \geq 2$,*

(1) $M_n(r)/\log \frac{1+r'}{r}$ *is an increasing function from $(0,1)$ onto $(1,\infty)$.*

(2) $M_n(r)/\log \frac{\lambda_n(1+r')}{2r}$ *is a decreasing function from $(0,1)$ onto $(0,1)$.*

(3) $M_n(r)^{n-1} \log \frac{1+r}{1-r}$ is an increasing function from $(0,1)$ onto $(0, 2A_n^{n-1})$, where

$$A_n = \left(\frac{\omega_{n-1}}{2\omega_{n-2}}\right)^{1/(n-1)} \int_0^{\pi/2} (\sin t)^{(2-n)/(n-1)} dt.$$

(4) For $n \geq 3$, $M_n(r)/\mu(r)$ is an increasing function from $(0,1)$ onto $(1, \infty)$.

From Theorem 2.13(4) it follows that $M_n(r) \geq \mu(r)$. If we combine this inequality with the inequalities (3.13)–(3.15) below, then we obtain an improved form of the lower bound in Corollary 2.12.

**2.14. Theorem.** For $n \geq 2$,

$$A_n(\frac{1}{2}\mu(\frac{1-r}{1+r}))^{1/(1-n)} \leq M_n(r) \leq A_n(\frac{1}{2} \log \frac{1+r}{1-r})^{1/(1-n)},$$

where $A_n$ is as in Theorem 2.13.

Several properties of $M_n(r)$ were studied in [AVV5]. This function satisfies certain non-linear functional inequalities, such as [AVV8, Theorem 2.9]:

$$(2.15) \qquad M_n\left(\frac{2s}{1+s^2}\right) + M_n\left(\frac{2t}{1+t^2}\right) \leq 2M_n\left(\frac{2\sqrt{st}}{1+st}\right), \quad n \geq 2, \quad s, t \in (0,1).$$

Note that equality holds here for $s = t$. Another such inequality appears in Theorem 3.16(2) below. From (2.6) it follows that $\gamma_n(t) = \omega_{n-1}M_n(1/t)^{1-n}$ for $t > 1$. It is well known [Vu2, Lemma 5.53] that

$$(2.16) \qquad \gamma_n(t) = 2^{n-1}\tau_n(t^2 - 1)$$

for $t > 1$. From these relations one can derive the properties of $\gamma_n(t)$ and $\tau_n(s)$ from those of $M_n(r)$. For $n = 2$ an explicit expression for $\mu(r) = M_2(r)$ is given by formula (3.11) below, and this allows some of the above results for $M_n(r)$ to be sharpened when $n = 2$, as we shall see in Section 3.

## 3. Special functions

The modulus of a curve family can be computed explicitly only in some special cases, such as curve families joining the boundary components of doubly-connected plane domains.

In particular, a ring domain in $\overline{\mathbf{R}}^2$ whose complement consists of two segments on the real axis can be mapped conformally onto an annulus. In this case the mapping is given by an elliptic function. Such explicit conformal maps account for the frequent occurrence of higher transcendental functions in the study of conformal invariants.

Important examples of such special functions are complete elliptic integrals of the first and second kind, both of which are special cases of the Gaussian hypergeometric function $F(a, b; c; r)$. The capacities $\gamma_2(s)$ and $\tau_2(t)$ of the plane Grötzsch and

Teichmüller rings, respectively, can be expressed in terms of $\mathcal{K}(r)$ (see (3.11) below). We recall the basic properties of these functions.

The Gaussian *hypergeometric function* is defined for $r \in (-1,1)$ by

$$(3.1) \qquad F(a,b;c;r) = {}_2F_1(a,b;c;r) = \sum_{n=0}^{\infty} \frac{(a,n)(b,n)}{(c,n)} \frac{r^n}{n!},$$

where $(a,n)$ denotes the ascending factorial function $(a,n) = a(a+1)\ldots(a+n-1)$, $n = 1,2,\ldots$ and $(a,0) = 1$ for $a \neq 0$. The sum is well-defined at least when $(c,n) \neq 0$, i.e. when $c \neq 0, -1, -2, \ldots$. Many elementary transcendental functions are special cases of $F(a,b;c;r)$; for an extensive list of such cases see [AS, PBM]. Two such special cases are the complete elliptic integrals

$$(3.2) \qquad \mathcal{K}(r) = \frac{\pi}{2}F\left(\frac{1}{2},\frac{1}{2};1;r^2\right), \quad \mathcal{E}(r) = \frac{\pi}{2}F(-\frac{1}{2},\frac{1}{2};1;r^2).$$

Equivalently, these functions can also be defined as integrals

$$(3.3) \qquad \mathcal{K}(r) = \int_0^{\pi/2} \frac{dt}{\sqrt{1-r^2\sin^2 t}}, \quad \mathcal{E}(r) = \int_0^{\pi/2} \sqrt{1-r^2\sin^2 t}\; dt.$$

It follows from (3.3) that $\mathcal{K}(0) = \mathcal{E}(0) = \pi/2$, $\mathcal{K}(1-) = \infty$, and $\mathcal{E}(1) = 1$. The following functional identities are due to Landen:

$$(3.4) \qquad \mathcal{K}\left(\frac{2\sqrt{r}}{1+r}\right) = (1+r)\mathcal{K}(r), \quad \mathcal{K}\left(\frac{1-r}{1+r}\right) = \frac{1+r}{2}\mathcal{K}(r'),$$

where $r' = \sqrt{1-r^2}$. The function $\mathcal{K}(r)$ occurs frequently in geometric function theory; see e.g. (3.11) below.

For $a, b > 0$ the *arithmetic-geometric mean* $\mathrm{ag}(a,b)$ of $a$ and $b$ is defined as the common limit of the sequences $(a_n)$ and $(b_n)$, with $a_1 = a$, $b_1 = b$, and

$$(3.5) \qquad a_{n+1} = (a_n + b_n)/2, \quad b_{n+1} = \sqrt{a_n b_n}.$$

Denote $\mathrm{ag}(1,r)$ by $\mathrm{ag}(r)$. Then $\sqrt{r} < \mathrm{ag}(r) < (1+r)/2$. Gauss proved the remarkable identity

$$(3.6) \qquad 2\mathrm{ag}(r')\mathcal{K}(r) = \pi, \quad r' = \sqrt{1-r^2}.$$

This identity provides a rapidly converging algorithm for numerical computation of $\mathcal{K}(r)$.

The Landen identities (3.4) have the following generalizations for $a, b \in (0,1)$ [AVV6]:

$$(3.7) \quad \begin{cases} \mathcal{K}\left(\dfrac{2\sqrt[4]{ab}}{\sqrt{(1+a)(1+b)}}\right) \le (1+\sqrt{ab})\mathcal{K}(\sqrt{ab}), \\[4mm] 2\mathcal{K}'\left(\dfrac{2\sqrt[4]{ab}}{\sqrt{(1+a)(1+b)}}\right) \ge (1+\sqrt{ab})\mathcal{K}'(\sqrt{ab}); \quad \mathcal{K}'(r) = \mathcal{K}(r'). \end{cases}$$

It follows from (3.4) that both inequalities in (3.7) reduce to identities when $a = b$.

The next theorem yields useful estimates for $\mathcal{K}(r)$ in terms of elementary functions. These inequalities are consequences of *monotone properties* of the function $\mathcal{K}(r)$.

**3.8. Theorem.** (1) *The function* $\mathcal{K}(r) + \log r'$ *is strictly decreasing from* $(0,1)$ *onto* $(\log 4, \pi/2)$ *and* $\mathcal{K}(r) + (\pi/4)\log r'$ *is strictly increasing from* $(0,1)$ *onto* $(\pi/2, \infty)$.

(2) *The function* $\mathcal{K}(r)/\log (4/r')$ *is strictly decreasing from* $(0,1)$ *onto* $(1, \frac{\pi}{\log 16})$ *and* $\mathcal{K}(r)/\log (e^2/r')$ *is strictly increasing from* $(0,1)$ *onto* $(\pi/4, 1)$.

From Theorem 3.8 one obtains e.g. the following inequalities [AVV6], for $0 < r < 1$:

$$(3.9) \quad 1 < \frac{\mathcal{K}(r)}{\log (4/r')} < \frac{\pi}{\log 16}, \quad \frac{\pi}{4} < \frac{\mathcal{K}(r)}{\log (e^2/r')} < 1, \quad \frac{\pi}{4} < \frac{\mathcal{K}(r) - \frac{\pi}{2}}{\log (1/r')} < 1.$$

These inequalities give sharp estimates for $\mathcal{K}(r)$. Moreover, by using the Landen identities (3.4) together with (3.9) we obtain new inequalities which have a different appearance from (3.9) and, perhaps, are sharper than (3.9). The lower bound in the following inequality was recently proved by R. Kühnau [Kü], while the upper bound remains an open problem [AVV6]:

$$(3.10) \quad \frac{9}{8+r^2} < \frac{\mathcal{K}(r)}{\log (4/r')} < \frac{9.2}{8+r^2}.$$

The conformal capacity $\gamma_2(s)$, $s > 1$, of the plane Grötzsch ring is given explicitly by

$$(3.11) \quad \gamma_2(s) = \frac{2\pi}{\mu(1/s)}; \quad \mu(r) = \frac{\pi\mathcal{K}(r')}{2\mathcal{K}(r)}.$$

The function $\mu(r)$ defined by (3.11) satisfies the functional identities

$$(3.12) \quad \mu(r)\mu(r') = \frac{\pi^2}{4}, \quad \mu(r) = 2\mu\left(\frac{2\sqrt{r}}{1+r}\right), \quad \mu(r)\mu\left(\frac{1-r}{1+r}\right) = \frac{\pi^2}{2}.$$

The first identity in (3.12) follows immediately from (3.11), and the other two are consequences of (3.4). Numerical computation of the values of $\mu(r)$ is efficiently accomplished by means of (3.6). Well-known bounds for $\mu(r)$ include [LV]

$$(3.13) \quad \log \frac{(1+\sqrt{r'})^2}{r} < \mu(r) < \log \frac{2(1+r')}{r}.$$

The latter upper bound has the drawback that it does not tend to zero as $r$ tends to unity. Note that $\mu(0+) = \infty$, $\mu(1-) = 0$. To rectify this anomaly we combine the inequality (3.13) with (3.12) and obtain e.g. the inequality

$$(3.14) \quad \max\left\{\log\frac{(1+\sqrt{r'})^2}{r}, \frac{\pi^2}{4\log\frac{2(1+r)}{r'}}\right\} < \mu(r) < \min\left\{\log\frac{2(1+r')}{r}, \frac{\pi^2}{4\log\frac{(1+\sqrt{r})^2}{r'}}\right\},$$

where now the majorant tends to zero as $r$ tends to unity. The inequalities (3.9) yield, for example,

$$(3.15) \quad \frac{\pi}{2}\frac{\log(4/r)}{\log(e^2/r')} < \mu(r) < \frac{\pi}{2}\frac{\log(e^2/r)}{\log(4/r')}.$$

Inequality (3.13) shows that the function $\mu(r)$ behaves like $\log(4/r)$ as $r$ tends to zero. The next theorem gives inequalities for $\mu(r)$ which show that $\mu(r)$ essentially transforms products into sums and thus behaves like a logarithm.

**3.16. Theorem.** *For* $a, b \in (0,1)$ *let* $c = \min\{a', b'\}$. *Then*

$$(1) \quad \mu\left(\frac{ab}{(1+c)^2}\right) \le \mu(a) + \mu(b) \le \mu\left(\frac{ab}{(1+a')(1+b')}\right) \le 2\mu(\sqrt{ab}),$$

*with equality for* $a = b$. *Further, for each* $n \ge 2$,

$$(2) \quad M_n\left(\frac{ab}{(1+c)^2}\right) \le M_n(a) + M_n(b) \le 2M_n(\sqrt{ab}).$$

The functions $\text{ag}(r)$, $\mathcal{K}(r)$, and $\mu(r)$ belong to the standard collection of special functions of geometric function theory. To study the action of quasiconformal mappings we need to augment this collection by functions tailored to this particular purpose. Several such functions look like a "perturbed identity". More precisely, we consider

$$(3.17) \quad \varphi_{K,n}(r) = M_n^{-1}(\alpha M_n(r)) = \frac{1}{\gamma_n^{-1}(K\gamma_n(1/r))}, \quad \alpha = K^{1/(1-n)},$$

for $K > 0$, $0 < r < 1$. The formula (3.17) yields immediately a fundamental composition property $\varphi_{AB,n}(r) = \varphi_{A,n}(\varphi_{B,n}(r))$ satisfied by $\varphi_{K,n}$. Since $\varphi_{1,n}(r) = r$, the parameter $K$ regulates the deviation of $\varphi_{K,n}(r)$ from the identity in the sense of Theorem 3.18 below. The function $\varphi_{K,n}(r)$ and its relatives arise naturally in the distortion theory of quasiconformal maps (see Section 4). Yet many properties of these functions are independent of quasiconformal maps. For instance, the next result follows easily from Theorem 2.14 [Vu2, 8.47] and from the monotone property in Theorem 2.7 [AVV1, AVV8].

**3.18. Theorem.** *For* $K \ge 1$, *let* $L = (\lambda_n/2)^{1-\alpha}$, $\ell = (\lambda_n/2)^{1-\beta}$, $\alpha = 1/\beta = K^{1/(1-n)}$, *and for* $r \in (0,1)$ *let* $A(r) = r/(1+r')$, $r' = \sqrt{1-r^2}$. *Then*

$$(1) \quad r^\alpha \le \varphi_{K,n}(r) \le \lambda_n^{1-\alpha} r^\alpha \le 2^{1-(1/K)} K r^\alpha,$$

$$(2) \qquad \varphi_{K,n}(r) \leq \text{th}(2\text{arth}(LA(r)^\alpha)), \text{ for } r \in (0, 2\ell/(1+\ell^2)),$$

$$(3) \qquad \varphi_{1/K,n}(r) \geq \text{th}(2\text{arth}(\ell A(r)^\beta)),$$

$$(4) \qquad \tau_n^{-1}(\tau_n(r)/K) \leq 4^{3-(1/K)} r^{1/K}, \text{ for } r \in (0, 2^{2-3K}).$$

It should be pointed out that Theorem 3.18(2) is new even for $n = 2$ and is a substantial improvement over the classical inequality $\varphi_{K,2}(r) \leq 4^{1-(1/K)} r^{1/K}$ (see 3.18(1) and [LV, (3.6), p. 65]).

For $n \geq 3$ there is no explicit formula such as (3.11), nor is a functional identity such as (3.12) known for $M_n(r)$. In the two-dimensional case we can translate the functional identities in (3.12) to the case of the function $\varphi_{K,2}(r)$, obtaining identities between $\varphi_{K,2}(r)$, and $\varphi_{1/K,2}(r')$. For instance, from the first identity in (3.12) we obtain a Pythagorean formula [AVV4]

$$(3.19) \qquad (\varphi_{K,2}(r))^2 + (\varphi_{1/K,2}(r'))^2 = 1.$$

Thus in some sense the non-linear functional inequalities (2.15) and 3.16(2) are substitutes for the functional identity (3.12) for the case $n \geq 3$. More such inequalities are given in [AVV5], e.g.

$$M_n(r) \leq 2M_n(\frac{2\sqrt{r}}{1+r}), \quad M_n(r)M_n(r') \geq \frac{\pi^2}{4},$$

of which the second inequality follows directly from Theorem 2.13(4) and (3.12). No analog of (3.19) is known for $\varphi_{K,n}(r)$ in higher dimensions $n \geq 3$ (see, however, Theorem 4.2(2)).

## 4. Distances and quasiconformal maps

One of the basic properties of quasiconformal maps is local Hölder continuity, which implies for instance that sets of zero $n-$measure are mapped onto sets of zero $n-$measure. On the other hand, these mappings are not, in general, locally bilipschitzian and thus e.g. the Hausdorff dimension is not invariant under these mappings.

The local Hölder continuity of quasiconformal maps in $\mathbf{R}^n$ was established in the early 1960's. Yet there are a number of open problems related to more refined aspects of Hölder continuity such as behavior of various constants as $K$ tends to unity. Most of these results make use of special functions such as $\varphi_{K,n}(r)$ and $\tau_n^{-1}(\tau_n(t)/K)$ studied in Section 3. Thus any new piece of information about these functions may yield information about such Hölder-continuity results. For the sake of simplicity we shall here consider only quasiconformal mappings, although some results hold even for the more general class of quasiregular mappings. For studies of the latter class we refer the reader to [BI, I, MRV, Re3, R, Vu1, Vu2].

The main point of this section is that many Hölder-continuity results or distortion theorems of the kind mentioned above can be derived in a unified way by use of two main ideas: (1) application of the quasi-invariance of the modulus (1.1) for the conformal invariants $\lambda_G(x,y)$ and $\mu_G(x,y)$ and (2) estimates of these invariants in terms of geometric quantities. Here we shall give some results of this type without full proofs; for more details the reader is referred to [Vu2, Sections 8 and 11].

For $K \geq 1, n \geq 2$, and $0 \leq r < 1$, we define

$$(4.1) \qquad \varphi^*_{K,n}(r) = \sup \{|f(x)| : f \in QC_K(B^n),\ f(0) = 0,\ |x| \leq r\},$$

where $QC_K(B^n)$ denotes the family of all $K$–quasiconformal mappings of $B^n$ with $f(B^n) \subset B^n = \{x \in \mathbf{R}^n : |x| < 1\}$. Recall the function $\varphi_{K,n}(r)$ defined in Section 3.

**4.2. Theorem.** *(Schwarz Lemma for quasiconformal maps). For $K \geq 1$, $n \geq 2$, $0 < r < 1$,*

$$(1) \qquad \qquad \varphi^*_{K,n}(r) \leq \varphi_{K,n}(r),$$

$$(2) \qquad \qquad (\varphi^*_{K,n}(r))^2 \leq 1 - (\varphi_{1/K,n}(r'))^2,$$

*where $r' = \sqrt{1 - r^2}$. Furthermore, equality holds in (1) and (2) for $n = 2$ for each $r \in (0, 1)$.*

The first part of Theorem 4.2 is due to O. Martio, S. Rickman, and J. Väisälä [MRV], whereas the second part is from [Vu1]. The equality statement follows from [LV, p.64] and (3.19). This result shows that quasiconformal mappings are Hölder continuous (cf. 3.18). Earlier similar or related results are due to E. D. Callender [Cal], F. W. Gehring [G2], and Yu. G. Reshetnyak [Re1, Re2]; see [Re3] and [Vu2, 11.50] for details.

A generic application of Theorem 4.2 is the following local Hölder-continuity theorem [AVV8].

**4.3. Theorem.** *For $f \in QC_K(B^n)$ and $x, y \in \overline{B}^n(r)$, $r \in (0, 1)$, $\alpha = K^{1/(1-n)}$,*

$$(1) \qquad \qquad |f(x) - f(y)| \leq a|x - y|^\alpha,$$

$$(2) \qquad \qquad |f(x) - f(y)| \leq b|x - y|^{1/K},$$

*where $a = \lambda_n^{1-\alpha}(1 - r^2)^{-\alpha} \leq 2^{1-(1/K)} \cdot K\,(1 - r^2)^{-\alpha}, b = \min\{2, K\}2^{1-1/K}(1 - r^2)^{-1/K}$.*

The proof of the first part of Theorem 4.2 uses the conformal invariant $\mu_{B^n}(x, y)$ together with an explicit formula for this invariant in terms of the hyperbolic distance $\rho(x, y)$ and the Grötzsch capacity [Vu2, 8.6(1)]. Similarly, the second part is based on the properties of $\lambda_{B^n}(x, y)$. In view of (3.19), for $n = 2$ the two bounds in Theorem 4.2 coincide, and it can be shown that they are different for $n \geq 3$ [Vu2, 7.58]. Note also that 3.18(1) together with 4.2(1) yields $\varphi^*_{K,n}(r) \leq \lambda_n^{1-\alpha}r^\alpha$, whereas [AVV5] gives $\varphi^*_{K,n}(r) \leq 2^{2-(1/K)}r^{1/K}$ for $0 < r < 1$. In fact, it can be shown [AVV2] that the sharp inequality

$$r^{-1/K}\varphi^*_{K,n}(r) \leq \min\{4^{1-(1/K^2)}, 8^{1-(1/K)}\}$$

holds for all $n \geq 2$ and all $r \in (0,1)$. Observe that the exponents $\alpha = K^{1/(1-n)}$ and $1/K$ are different for $n \geq 3$ and $K > 1$.

We next derive a similar result for distortion in the spherical chordal metric $q(x,y)$ by using the conformal invariant $\lambda_G(x,y)$. Recall that the chordal metric in $\overline{\mathbf{R}}^n$ is given by

$$q(x,y) = \frac{|x-y|}{\sqrt{(1+|x|^2)(1+|y|^2)}}, \quad q(x,\infty) = 1/\sqrt{1+|x|^2}$$

for $x, y \in \mathbf{R}^n$, and that the absolute cross ratio is given by

$$|a,b,c,d| = \frac{q(a,c)q(b,d)}{q(a,b)q(c,d)}$$

whenever $a, b, c, d$ are distinct points in $\overline{\mathbf{R}}^n$. For $E \subset \overline{\mathbf{R}}^n$, $F \subset \mathbf{R}^n$, and $x \in \mathbf{R}^n$ we denote by $q(E)$ and $d(x,F)$ the spherical diameter of $E$ and the euclidean distance from $x$ to $F$, respectively.

**4.4. Lemma [Vu4].** *For each $x \in \mathbf{R}^n \setminus [0, e_1]$ there exist continua $E$ and $F$ with $0, e_1 \in E$ and $x, \infty \in F$ such that*

$$M(\Delta(E,F;\mathbf{R}^n)) \leq \tau_n\left(\frac{|x| + |x - e_1| - 1}{2}\right).$$

*Furthermore, equality holds if $x = te_1, t > 1$ .*

**Proof.** Let $h$ be a Möbius transformation of $\mathbf{R}^n$ taking $x, 0, e_1, \infty$ onto $-e_1, -y, y, e_1$, respectively, where $|y| < 1$. With $E_1 = [-|y|e_1, |y|e_1], E' = [-y, y], F' = [-e_1, \infty] \cup [e_1, \infty], E = h^{-1}(E')$, and $F = h^{-1}(F')$, we have $M(\Delta(E', F')) \leq M(\Delta(E_1, F'))$ and

$$p(x) \leq M(\Delta(E,F)) \leq M(\Delta(E_1, F')) = \tau_n\left(\frac{1-|y|^2}{4|y|}\right) = \tau_n\left(\frac{|x| + |x - e_1| - 1}{2}\right). \quad \square$$

There is a natural extremal problem associated with Lemma 4.4 as follows: Find the infimum of the modulus on the left side in the above inequality, where the infimum is taken over all pairs of continua $E$ and $F$ with the stated properties. This extremal problem was first considered by O. Teichmüller [T, p.169] in the case where $n = 2$. A few years later M. Schiffer [S] proved that in this particular case the infimum is in fact a minimum and that the minimizing continua are the images under an elliptic function of the boundary components of a ring domain. Even the simple sets given in Lemma 4.4 are nearly extremal continua for this problem in all dimensions $n \geq 2$.

**4.5. Corollary.** *Let $G = \mathbf{R}^n \setminus \{0\}$ and $x, y \in G$ with $x \neq y$. Then*

$$\lambda_G(x,y) \leq \tau_n\left(\frac{|x-y| + ||x| - |y||}{2\min\{|x|,|y|\}}\right) \leq \tau_n\left(\frac{|x-y|}{2\min\{|x|,|y|\}}\right)$$

$$= \tau_n(\max\{|x,0,y,\infty|, |y,0,x,\infty|\}/2).$$

**Proof.** We may assume that $y = e_1$ and $|x| \geq 1$. We join $y$ to $0$ and $x$ to $\infty$ by the sets $E$ and $F$ constructed in Lemma 4.4. The result follows from Lemma 4.4. $\square$

**4.6. Theorem.** *Let $G$ be a proper subdomain of $\mathbf{R}^n$ and let $f : G \to f(G)$ be a $K$-quasiconformal mapping of $G$ onto a subdomain of $\mathbf{R}^n$. Then, for $x, y \in G$,*

$$q(f(x), f(y))q(\partial f(G)) \leq 128(|x - y|/d(x, \partial G))^{1/K}.$$

**Proof.** There is nothing to prove if $x = y$. Hence we may assume that $x \neq y$ and fix $a, d \in \partial f(G)$ with $q(a, d) = q(\partial f(G)) > 0$. Let $r = |x - y|/d(x, \partial G)$. If $r \geq 2^{-7K}$ the right hand side is at least one and there is nothing to prove. For $r \in (0, 2^{-7K})$ we argue as follows. With $B_x = B^n(x, d(x, \partial G))$ we have by [Vu2, Lemma 8.8 (1)]

$$\lambda_G(x, y) \geq \lambda_{B_x}(x, y) = \frac{1}{2}\tau_n\left(\frac{r^2}{1 - r^2}\right).$$

Let $D = \overline{\mathbf{R}}^n \setminus \{a, d\}$. By Corollary 4.5 we have

$$\lambda_{f(G)}(f(x), f(y)) \leq \lambda_D(f(x), f(y))$$

$$\leq \tau_n(\max\{|f(x), a, f(y), d|, |f(y), a, f(x), d|\}/2)$$

$$\leq \tau_n(q(f(x), f(y))\, q(\partial f(G))/2),$$

where the last estimate is crude. Since $\lambda_G(x, y) \leq K\lambda_{f(G)}(f(x), f(y))$ the above inequalities yield

$$q(f(x), f(y))\, q(\partial f(G)) \leq 2\tau_n^{-1}\left(\frac{1}{2K}\tau_n\left(\frac{r^2}{1 - r^2}\right)\right).$$

Now $r < 2^{-7K}$ implies $r^2/(1 - r^2) < 2^{-7K}$ and by Theorem 3.18 (4) we get

$$q(f(x), f(y))\, q(\partial f(G)) \leq 2\cdot 4^{3-(1/2K)}\left(\frac{r^2}{1 - r^2}\right)^{1/(2K)} \leq 128\left(\frac{r}{2\sqrt{1 - r^2}}\right)^{1/K} \leq 128r^{1/K},$$

as desired. □

F. W. Gehring [G3, p.72] earlier proved a result similar to Theorem 4.6 by a different method with a constant depending on the dimension $n$ in place of the universal constant 128. "Dimension-free" results similar to Theorems 4.3 and 4.6 were first proved in [Vu1, AVV1, Vu2]. Very recently, J. Väisälä [V3] has begun a study of quasiconformal maps in infinite-dimensional Banach spaces.

We next give another application of Lemma 4.4, this time for the two-dimensional case, where we have the following sharp result (see [Vu4]).

**4.7. Theorem.** *Let $E$ be a continuum with $0, e_1 \in E$ and let $f : \overline{\mathbf{R}}^2 \setminus E \to \overline{\mathbf{R}}^2 \setminus \overline{B}^2$ be a conformal mapping onto the complement of the closed unit disk with $f(\infty) = \infty$. Then for $x \in \mathbf{R}^2 \setminus E$*

$$|f(x)| \leq (\sqrt{1 + m} + \sqrt{m})^2,$$

where $m = \min\{|x|, |x - e_1|\}$. *Conversely, for $x \in \mathbf{R}^2 \setminus [0, e_1]$ there exists a circular arc $E = E_x$ and a conformal mapping $f : \overline{\mathbf{R}}^2 \setminus E \to \overline{\mathbf{R}}^2 \setminus \overline{B}^2$ with $f(\infty) = \infty$ such that*

$$|f(x)| \geq u + \sqrt{u^2 - 1}, \quad u = |x| + |x - e_1|.$$

*For $x = te_1$, $t > 1$, both bounds reduce to $(\sqrt{t} + \sqrt{t-1})^2$ and hence the result is sharp.*

A common feature in the proofs of the results in this section is the use of quasi-invariance properties

$$(4.8) \qquad \mu_{fD}(f(x), f(y)) \leq K\mu_D(x, y), \quad \lambda_D(x, y) \leq K\lambda_{fD}(f(x), f(y)),$$

under a $K$-quasiconformal mapping $f$ of $D$ onto $fD$. Recall that by Remark 2.4(2) $\mu_D(x, y)$ and $\lambda_D(x, y)^{-1/n}$ define conformally invariant metrics on $D$. Thus the inequalities in (4.8) define conformally invariant Lipschitz classes. In [LF] J. Ferrand raised the question whether the second inequality in (4.8) alone implies that the homeomorphism $f$ is quasiconformal in $D$. A non-quasiconformal homeomorphism $g = (g_1, g_2)$ of a plane Jordan domain that satisfies the second inequality in (4.8) appears in [FMV]. This happens because the coordinate functions $g_1, g_2$ are Hölder continuous with different exponents in a neighborhood of a certain point $x_1$, so that the linear dilatation of $g$ is infinite at $x_1$. Besides being Hölder continuous, the mappings satisfying one of the inequalities in (4.8) share some other properties of quasiconformal mappings. It appears that few results exist in the literature concerning homeomorphisms, between domains in $\mathbf{R}^n$, whose coordinate functions are Hölder continuous, possibly with different exponents.

4.9. **Remarks.** We have already pointed out at the end of Section 3 that no analog of the identity (3.19) is known for $\varphi_{K,n}(r), n \geq 3$. In some sense, the inequality in Theorem 4.2(2) is an analog of (3.19) involving $\varphi_{K,n}^*(r)$. The function $\varphi_{K,n}^*(r)$ also satisfies some other natural inequalities, as shown in [AVV3]. One of these is

$$(4.10) \qquad \varphi_{K,n}^*(B(r, s)) \leq B(\varphi_{K,n}^*(r), \varphi_{K,n}^*(s)); \quad B(r, s) = \frac{r + s}{1 + rs},$$

for $r, s \in (0, 1), K \geq 1$, with equality for $K = 1$. The above inequality also shows that $\varphi_{K,n}^*(\text{th}(\rho(x, y)/2))$ defines a conformally invariant metric on $B^n$, where $\rho(x, y)$ denotes the hyperbolic distance between $x, y \in B^n$. Further, the inequality

$$(4.11) \qquad \varphi_{K,n}^*(r) \leq \text{th}(\text{arth}(r) + (\beta - 1)M_n(r')), \quad \beta = K^{1/(n-1)},$$

for $K \geq 1$ is given in [AVV8].

## 5. Quadruples and quasiconformal maps

Möbius transformations are characterized by the fact that they preserve the absolute cross ratio. There is an analogous result for $K$-quasiconformal maps of $\mathbf{R}^n$ (see Theorem 5.1 and (5.2) below). As a corollary one may prove that angles are almost preserved under quasiconformal maps and may obtain other results such as a version of Mori's classical theorem.

Let $\mathcal{F}(n, K)$ denote the family of all $K$−quasiconformal mappings $f : \overline{\mathbf{R}}^n \to \overline{\mathbf{R}}^n$ normalized by $f(0) = 0$, $f(e_1) = e_1$, and $f(\infty) = \infty$. For $t > 0$ let

$$\eta_{K,n}(t) = \sup \{|f(x)| : |x| \leq t, \ f \in \mathcal{F}(n, K)\}.$$

The following result [Vu3] (see also [AVV9]) is basic:

**5.1. Theorem.** *For $K \geq 1$, $n \geq 2$, we have*

(1)
$$\eta_{K,n}(1) \leq \inf_{0 < t < 1} \frac{A_{K,n}(t)}{B_{K,n}(t)},$$

*where*

$$A_{K,n}(t) = \frac{\varphi_{K,n}^2(\sqrt{t})}{1 - \varphi_{K,n}^2(\sqrt{t})}, \quad B_{K,n}(t) = \varphi_{1/K,n}^2\left(\sqrt{\frac{t}{1+t}}\right) = A_{K,n}^{-1}(t).$$

*Further,*

(2)
$$\eta_{K,n}(1) \leq \exp\left(6(K+1)^2\sqrt{K-1}\right),$$

(3)
$$\begin{cases} \eta_{K,n}(t) \leq \eta_{K,n}(1)\varphi_{K,n}(t), & 0 < t \leq 1, \\ \eta_{K,n}(t) \leq \eta_{K,n}(1)/\varphi_{1/K,n}(1/t), & t \geq 1. \end{cases}$$

It follows from the above results that a $K$−quasiconformal mapping $f : \overline{\mathbf{R}}^n \to \overline{\mathbf{R}}^n$ satisfies the double inequality

(5.2)
$$\frac{1}{\eta_{K,n}(1/s)} \leq |f(a), f(b), f(c), f(d)| \leq \eta_{K,n}(s), \quad s = |a, b, c, d|,$$

for all distinct points $a, b, c, d \in \overline{\mathbf{R}}^n$. Since $\eta_{1,n}(t) = t$ we see that Theorem 5.1, together with (5.2), implies the invariance of the absolute ratio under Möbius transformations when $K = 1$, and for each $K > 1$ we obtain a sharp bound for the variation of the absolute cross ratio under a $K$−quasiconformal mapping. Further, we can deduce the following result from Theorems 5.1 and 3.18.

**5.3. Theorem.** *Let $f : \overline{\mathbf{R}}^n \to \overline{\mathbf{R}}^n$ be a $K$−quasiconformal mapping, let $x, y, z, w$ be distinct points in $\overline{\mathbf{R}}^n$, and let $s_1 = |x, z, y, w|$, $s_2 = |x, z, w, y|$, $s_1' = |f(x), f(z), f(y), f(w)|$, $s_2' = |f(x), f(z), f(w), f(y)|$. Then*

$$c_1(s_1 + s_2)^\alpha \leq s_1' + s_2' \leq c_2(s_1 + s_2)^\beta,$$

*where $\alpha = K^{1/(1-n)} = 1/\beta$, $c_1 = 2^{1-(\beta/\alpha)}\lambda_n^{1-\beta}/\eta_{K,n}(1)$, $c_2 = 2^{1-(\beta/\alpha)}\lambda_n^{\beta-1}\eta_{K,n}(1)$.*

Since

$$|x, z, y, \infty| + |x, z, \infty, y| = \frac{|x-y| + |y-z|}{|x-z|},$$

Theorem 5.3 implies

**5.4. Corollary.** *Let $f : \overline{\mathbf{R}}^n \to \overline{\mathbf{R}}^n$ be a $K$–quasiconformal map with $f(\infty) = \infty$ and let $x, y, z$ be distinct points in $\mathbf{R}^n$. Then*

$$c_3 \left( \frac{|x-z|}{|x-y|+|y-z|} \right)^\beta \leq \frac{|f(x)-f(z)|}{|f(x)-f(y)|+|f(y)-f(z)|} \leq c_4 \left( \frac{|x-z|}{|x-y|+|y-z|} \right)^\alpha,$$

*where $c_3 = 1/c_2$, $c_4 = 1/c_1$, and $c_1, c_2$ are as in Theorem 5.3.*

Note that in view of Theorem 5.1(2) the constants $c_j$, $j = 1, \ldots, 4$, in 5.3 and 5.4 are all equal to 1 for $K = 1$.

**5.5. Theorem.** *Let $f : \overline{\mathbf{R}}^n \to \overline{\mathbf{R}}^n$ be a $K$–quasiconformal map with $f(\infty) = \infty$, $f(0) = 0$, and $f(B^n) \subset B^n$. Then*

$$|f(x) - f(y)| \leq C(n, K)|x - y|^\alpha, \quad \alpha = K^{1/(1-n)},$$

*for $x, y \in B^n$, where $C(n, K) = (2\lambda_n)^{1-\alpha} c_4$ and $c_4$ is as in Corollary 5.4.*

**Proof.** Corollary 5.4 and Theorem 3.18(1) yield

$$|f(x) - f(y)| \leq c_4(|f(x)| + |f(y)|) \left( \frac{|x-y|}{|x|+|y|} \right)^\alpha$$

$$\leq c_4 \lambda_n^{1-\alpha}(|x|^\alpha + |y|^\alpha) \frac{|x-y|^\alpha}{2^{\alpha-1}(|x|^\alpha + |y|^\alpha)}$$

$$= (2\lambda_n)^{1-\alpha} c_4 |x - y|^\alpha. \quad \square$$

Next, we recall a sharp angle-distortion result, in the plane, of S. Agard and F. W. Gehring [AG, Theorem 2], and then give an application.

**5.6. Theorem.** *Suppose that $f$ is a $K$–quasiconformal mapping of the extended plane with $f(\infty) = \infty$. Then for each triple of distinct finite points $z_1, z_0, z_2$,*

$$\varphi_{1/K,2}(\sin \frac{\alpha}{2}) \leq \sin \frac{\beta}{2} \leq \varphi_{K,2}(\sin \frac{\alpha}{2}),$$

*where*

$$\alpha = \arcsin \left( \frac{|z_1 - z_2|}{|z_1 - z_0| + |z_2 - z_0|} \right), \quad \beta = \arcsin \left( \frac{|f(z_1) - f(z_2)|}{|f(z_1) - f(z_0)| + |f(z_2) - f(z_0)|} \right).$$

**5.7. Corollary.** *Under the hypotheses of Theorem 5.6,*

$$(4\delta)^{1-K}(\sin \alpha)^K \leq \sin \beta \leq (4\delta)^{1-(1/K)}(\sin \alpha)^{1/K},$$

*where $\delta = \cos \left( \frac{\alpha}{2} \right) + \cos^2 \left( \frac{\alpha}{2} \right)$.*

**Proof.** If $\beta \in (0, \alpha]$ the second inequality is obvious. Hence we may assume that $\beta > \alpha$. Now by [AVV8, Theorem 3.1(3)],

$$\sin\,\beta = 2\sin\left(\frac{\beta}{2}\right)\cos\left(\frac{\beta}{2}\right) \leq 2\varphi_{K,2}(\sin\left(\frac{\alpha}{2}\right))\cos\left(\frac{\alpha}{2}\right)$$

$$\leq 2 \cdot 2^{1-(1/K)}(1 + \cos\left(\frac{\alpha}{2}\right))^{1-(1/K)}(\sin\left(\frac{\alpha}{2}\right))^{1/K}\cos\left(\frac{\alpha}{2}\right))$$

$$= (4\delta)^{1-(1/K)}(\sin\,\alpha)^{1/K},$$

which proves the second inequality. The proof of the first inequality is similar. □

**5.8. Theorem.** *Let* $f \in QC_K(B^2)$ *with* $f(0) = 0$ *and* $f(B^2) = B^2$. *Then for* $x, y \in B^2$

$$|f(x) - f(y)| \leq (8\delta)^{1-(1/K)}r|x - y|^{1/K} \leq 64^{1-(1/K)}|x - y|^{1/K},$$

*where*

$$\delta = \cos\left(\frac{\alpha}{2}\right) + \cos^2\left(\frac{\alpha}{2}\right), \quad \sin\,\alpha = \frac{|x - y|}{|x| + |y|},$$

$$r = \frac{|x|^{1/K} + |y|^{1/K}}{(|x| + |y|)^{1/K}}(\max\,\{1 + |x'|, 1 + |y'|\})^{1-1/K}.$$

**Proof.** By reflection [LV, p. 47] $f$ can be extended to a $K$–quasiconformal self-mapping of $\overline{\mathsf{R}}^2$ with $f(\infty) = \infty$. By [BB, Ch. 1, Section 16] $r \leq 4^{1-1/K}$. Hence Corollary 5.7 and [AVV8, Theorem 3.1(3)] together yield

$$|f(x) - f(y)| \leq (|f(x)| + |f(y)|)(4\delta)^{1-(1/K)}(\sin\,\alpha)^{1/K}$$

$$\leq (8\delta)^{1-1/K}(\sin\,\alpha)^{1/K}[(1 + |x'|)^{1-(1/K)}|x|^{1/K} + (1 + |y'|)^{1-(1/K)}|y|^{1/K}]$$

$$\leq (8\delta)^{1-(1/K)}r|x - y|^{1/K} \leq 64^{1-(1/K)}|x - y|^{1/K}. \quad □$$

**5.9. Remarks.** (1) A classical theorem of A. Mori yields a result similar to Theorem 5.5 with constant 16 in place of $C(2, K)$ for $n = 2$. By Theorem 5.8 the constant $C(2, K)$ can also be replaced by $c^{1-(1/K)}$, where $16 \leq c \leq 64$; see [AVV8] for references. In [LV, p. 68] it is shown that $C(2, K) \geq 16^{1-1/K}$.

(2) For recent results related to Theorem 5.5 see [AV; BP; FV; Vu2, p. 150; AVV8].

(3) For $n = 2$ one can replace the inequalities in Theorem 5.1 by the identity

$$\eta_{K,2}(r) = \frac{u^2}{1 - u^2}, \quad u = \varphi_{K,2}\left(\sqrt{\frac{r}{1 + r}}\right),$$

in which $u$ can be estimated with the aid of Theorem 3.18.

# References

[AS]  M. Abramowitz and I. A. Stegun, editors, Handbook of Mathematical Functions with Formulas, Graphs and Mathematical Tables, Dover, New York, 1965.

[AG]  S. Agard and F. W. Gehring, Angles and quasiconformal mappings, Proc. London Math. Soc. (3) 14A (1965), 1–21.

[AhB]  L. Ahlfors and A. Beurling, Conformal invariants and function–theoretic null-sets, Acta Math. 83 (1950), 101–129.

[A]  G. D. Anderson, Extremal rings in $n$–space for fixed and varying $n$, Ann. Acad. Sci. Fenn. A I 575 (1974), 1–21.

[AF]  G. D. Anderson and J. S. Frame, Numerical estimates for a Grötzsch ring constant, Constr. Approx. 4 (1988), 223–242.

[AV]  G. D. Anderson and M. K. Vamanamurthy, Hölder continuity of quasiconformal mappings of the unit ball, Proc. Amer. Math. Soc. 104 (1988), 227–230.

[AVV1]  G. D. Anderson, M. K. Vamanamurthy, and M. Vuorinen, Dimension-free quasiconformal distortion in $n$–space, Trans. Amer. Math. Soc. 297 (1986), 687–706.

[AVV2]  G. D. Anderson, M. K. Vamanamurthy, and M. Vuorinen, Sharp distortion theorems for quasiconformal mappings, Trans. Amer. Math. Soc. 305 (1988), 95–111.

[AVV3]  G. D. Anderson, M. K. Vamanamurthy, and M. Vuorinen, Inequalities for the extremal distortion function, Proc. of the 13th Rolf Nevanlinna Colloquium, Lecture Notes in Math. Vol. 1351, 1–11, Springer– Verlag, 1988.

[AVV4]  G. D. Anderson, M. K. Vamanamurthy, and M. Vuorinen: Distortion functions for plane quasiconformal mappings, Israel J. Math. 62 (1988), 1–16.

[AVV5]  G. D. Anderson, M. K. Vamanamurthy, and M. Vuorinen, Special functions of quasiconformal theory, Exposition. Math. 7 (1989), 97–138.

[AVV6]  G. D. Anderson, M. K. Vamanamurthy, and M. Vuorinen, Functional inequalities for complete elliptic integrals and their ratios, SIAM J. Math. Anal. 21 (1990), 536–549.

[AVV7]  G. D. Anderson, M. K. Vamanamurthy, and M. Vuorinen, Functional inequalities for hypergeometric functions and complete elliptic integrals, SIAM J. Math. Anal. 23 (1992) (to appear).

[AVV8]  G. D. Anderson, M. K. Vamanamurthy, and M. Vuorinen, Inequalities for quasiconformal mappings in the plane and in space (in preparation).

[BB]  E. F. Beckenbach and R. Bellman, Inequalities, Ergebnisse der Mathematik, Heft 30, Springer-Verlag, 1961.

[BP]  J. Becker and Ch. Pommerenke, Hölder continuity of conformal maps with quasiconformal extension, Complex Variables Theory Appl. 10 (1988), 267–272.

[BI]  B. Bojarski and T. Iwaniec, Analytical foundations of the theory of quasiconformal mappings in $R^n$, Ann. Acad. Sci. Fenn. Ser. A I Math. 8 (1983), 257–324.

[Cal]  E. D. Callender, Hölder continuity of $n$- dimensional quasiconformal mappings, Pacific J. Math. 10 (1960), 499–515.

[Car]   P. Caraman, n-Dimensional Quasiconformal (QCf) Mappings, Abacus Press, Tunbridge Wells, Kent, England, 1974.

[FV]    R. Fehlmann and M. Vuorinen, Mori's theorem for $n-$dimensional quasiconformal mappings, Ann. Acad. Sci. Fenn. Ser. A I 13 (1988), 111–124.

[FMV]   J. Ferrand, G. Martin, and M. Vuorinen, Lipschitz conditions in conformally invariant metrics, J. Analyse Math. 56 (1991), 187- -210.

[F]     B. Fuglede, Extremal length and functional completion, Acta Math. 98 (1957), 171–219.

[G1]    F. W. Gehring, Symmetrization of rings in space, Trans. Amer. Math. Soc. 101 (1961), 499–519.

[G2]    F. W. Gehring, Rings and quasiconformal mappings in space, Trans. Amer. Math. Soc. 103 (1962), 353–393.

[G3]    F. W. Gehring, Topics in quasiconformal mappings, Proc. Internat. Congr. Math. (Berkeley, California 1986), Vol. 1, 62–80, AMS, 1987.

[GV]    F. W. Gehring and J. Väisälä, The coefficients of quasiconformality of domains in space, Acta Math. 114 (1965), 1–70.

[Gr]    H. Grötzsch, Über die Verzerrung bei schlichten nichtkonformen Abbildungen und über eine damit zusammenhängende Erweiterung des Picardschen Satzes, Ber. Verh. Sächs. Akad. Wiss. Leipzig 80 (1928), 503–507.

[H]     J. Hersch, Longeurs extrémales et théorie des fonctions, Comment. Math. Helv. 29 (1955), 301–337.

[I]     T. Iwaniec, Some aspects of partial differential equations and quasiregular mappings, Proc. Internat. Congr. Math. (Warsaw, 1983), Vol. 2, 1193–1208, PWN, Warsaw, 1984.

[J]     J. A. Jenkins, The method of the extremal metric, Proc. Congr. on the Occasion of the solution of the Bieberbach conjecture, Amer. Math. Soc. 1986, 95–104.

[Ke]    J. A. Kelingos, Characterization of quasiconformal mappings in terms of harmonic and hyperbolic measure, Ann. Acad. Sci. Fenn. Ser. A I, No. 368 (1965), 1–16.

[Kr]    M. Kreines, Sur une classe de fonctions de plusieurs variables, Mat. Sbornik 9 (1941), 713–719.

[Kü]    R. Kühnau, Eine Methode, die Positivität einer Funktion zu prüfen, Zeitschrift f. angew. Math. u. Mech. (to appear).

[La]    M. Lavrentiev, Sur un critère différentiel des transformations homéomorphes des domaines à trois dimensions, Dokl. Akad. Nauk SSSR 22 (1938), 241–242.

[LV]    O. Lehto and K. I. Virtanen, Quasiconformal Mappings in the Plane, 2nd ed., Die Grundlehren der math. Wissenschaften, Band 126, Springer–Verlag, New–York–Heidelberg–Berlin, 1973.

[LF]    J. Lelong–Ferrand, Invariants conformes globaux sur les varictes riemanniennes, J. Differential Geom. 8 (1973), 487–510.

[Lo]    C. Loewner, On the conformal capacity in space, J. Math. Mech. 8 (1959), 411–414.

[Ma]    A. Markushevich, Sur certaines classes de transformations continues, Dokl. Akad. Nauk SSSR 28 (1940), 301–304.

[MRV] O. Martio, S. Rickman, and J. Väisälä, Distortion and singularities of quasiregular mappings, Ann. Acad. Sci. Fenn. Ser. A I 465 (1970), 1–13.

[PBM] A. P. Prudnikov, Yu. A. Brychkov, and O. Marichev, Integrals and Series, Vol. 3: More Special Functions, transl. from the Russian by G. G. Gould, Gordon and Breach Science Publishers, New York, 1988.

[Re1] Yu. G. Reshetnyak, A sufficient condition for Hölder continuity of a mapping, Soviet Math. Doklady 1 (1960), 76–78.

[Re2] Yu. G. Reshetnyak, Bounds on moduli of continuity for certain mappings, Siberian Math. J. 7 (1966), 879–886.

[Re3] Yu. G. Reshetnyak, Space Mappings with Bounded Distortion, Translations of Mathematical Monographs Vol. 73, Amer. Math. Soc. Providence, R.I., 1989.

[R] S. Rickman, Quasiregular Mappings (to appear).

[S] M. Schiffer, On the modulus of doubly-connected domains, Quart. J. Math. Oxford, Ser. 17 (1946), 197–213.

[Sh] B. V. Shabat, On the theory of quasiconformal mappings in space, Soviet Math. 1 (1960), 730–733.

[T] O. Teichmüller, Untersuchungen über konforme und quasikonforme Abbildung, Deutsche Math. 3 (1938), 621–678.

[V1] J. Väisälä, On quasiconformal mappings in space, Ann. Acad. Sci. Fenn. Ser. A I 298 (1961), 1–36.

[V2] J. Väisälä, Lectures on $n$–Dimensional Quasiconformal Mappings, Lecture Notes in Math. Vol. 229, Springer–Verlag, Berlin–Heidelberg–New York, 1971.

[V3] J. Väisälä, Free quasiconformality in Banach spaces I, Ann. Acad. Sci. Fenn. Ser. A I 15 (1990), 355–379.

[Vu1] M. Vuorinen, Conformal invariants and quasiregular mappings, J. Analyse Math. 45 (1985), 69–115.

[Vu2] M. Vuorinen, Conformal Geometry and Quasiregular Mappings, Lecture Notes in Math. Vol. 1319, Springer–Verlag, Berlin–Heidelberg–New York, 1988.

[Vu3] M. Vuorinen, Quadruples and spatial quasiconformal mappings, Math. Z. 205 (1990), 617–628.

[Vu4] M. Vuorinen, Conformally invariant extremal problems and quasiconformal mappings, Quart. J. Math. Oxford (to appear).

Quasiconformal Space Mappings
– A collection of surveys 1960–1990
Springer–Verlag (1992), 20–38
Lecture Notes in Mathematics Vol. 1508

# Topics in Quasiconformal Mappings

## F. W. GEHRING

### I. Introduction.

1. *Notation.* For $n \geq 1$ we let $\mathbf{R}^n$ denote euclidean $n$-space, and for $x \in \mathbf{R}^n$ and $0 < r < \infty$ we let $\mathbf{B}^n(x, r)$ denote the open $n$-ball with center $x$ and radius $r$, $\mathbf{S}^{n-1}(x, r) = \partial \mathbf{B}^n(x, r)$, $\mathbf{B}^n = \mathbf{B}^n(0, 1)$, and $\mathbf{S}^{n-1} = \mathbf{S}^{n-1}(0, 1)$. We also denote by $\overline{\mathbf{R}}^n = \mathbf{R}^n \cup \{\infty\}$ the one point compactification of $\mathbf{R}^n$ equipped with the chordal metric

$$q(x, y) = |p(x) - p(y)|, \tag{1.1}$$

where $p$ denotes stereographic projection of $\overline{\mathbf{R}}^n$ onto the sphere $\mathbf{S}^n$ in $\mathbf{R}^{n+1}$. Throughout this paper, all notions of topology and convergence will be taken with respect to this metric.

Suppose that $D$ and $D'$ are domains in $\overline{\mathbf{R}}^n$ and that $f: D \to D'$ is a homeomorphism. We let

$$H_f(x) = \limsup_{r \to 0} H_f(x, r) \tag{1.2}$$

for $x \in D \setminus \{\infty, f^{-1}(\infty)\}$, where for $0 < r < \text{dist}(x, \partial D)$

$$H_f(x, r) = \frac{\max\{|f(x) - f(y)| : |x - y| = r\}}{\min\{|f(x) - f(z)| : |x - z| = r\}}, \tag{1.3}$$

and we extend $H_f(x)$ to the points $\infty$ and $f^{-1}(\infty)$ by setting $H_f(\infty) = H_{f \circ g}(0)$ and $H_f(f^{-1}(\infty)) = H_{g \circ f}(f^{-1}(\infty))$, where $g(x) = x/|x|^2$. When $n \geq 2$, we call

$$K(f) = \begin{cases} \infty & \text{if } \sup_{x \in D} H_f(x) = \infty, \\ \text{ess sup}_{x \in D} H_f(x) & \text{if } \sup_{x \in D} H_f(x) < \infty \end{cases} \tag{1.4}$$

the *linear dilatation* of $f$ in $D$. For the purposes of this lecture, we say that $f$ is *quasiconformal* if $K(f) < \infty$ and $K$-*quasiconformal* if $K(f) \leq K$, $1 \leq K < \infty$. Thus a homeomorphism is quasiconformal if it distorts the shape of an infinitesimal $(n - 1)$-sphere about each point by at most a bounded factor; it is $K$-quasiconformal if, in addition, this factor does not exceed $K$ at almost every point.

The following result shows that the class of quasiconformal mappings is, as the name suggests, an extension of the family of conformal mappings.

---

This research was supported in part by grants from the National Science Foundation.

1.5. THEOREM. *Suppose that $D, D'$ are domains in $\overline{\mathbf{R}}^n$ and that $f: D \to D'$ is a homeomorphism. If $n = 2$, then $f$ is 1-quasiconformal if and only if $f$ or its complex conjugate is a meromorphic function of a complex variable in $D$. If $n \geq 3$, then $f$ is 1-quasiconformal if and only if $f$ is the restriction to $D$ of a Möbius transformation, i.e., the composition of a finite number of reflections in $(n-1)$-spheres and planes.*

When $n = 2$, Theorem 1.5 is simply a restatement of a theorem due to Menchoff [M3]. When $n \geq 3$, Theorem 1.5 is an extension of a well-known result of Liouville to a context which requires *no a priori* hypotheses on the smoothness of $f$ [G3, R3].

2. *Historical remarks.* Plane quasiconformal mappings have been studied for almost sixty years. They appear in the late 1920s in papers by Gröztsch, who considered the problem of determining the *most nearly conformal* homeomorphisms between pairs of topologically equivalent plane configurations with one conformal invariant [G14]. They occur later under the name *quasiconformal* in a paper by Ahlfors on covering surfaces [A1].

In the late 1930s Teichmüller vastly extended the study of Grötzsch to mappings between closed Riemann surfaces and obtained a very natural parameter space for surfaces of fixed genus $g$, a space which is homeomorphic to $\mathbf{R}^{6g-6}$ [T1]. At about the same time, Lavrentieff and Morrey generalized a classical result due to Gauss on the existence of isothermal coordinates by establishing versions of what is now known as the *measurable Riemann mapping theorem* for quasiconformal mappings [L1, M4].

In recent years, Ahlfors, Bers, and their school have greatly expanded the results of Teichmüller and applied plane quasiconformal mappings with success to a variety of areas in complex analysis, including kleinian groups and surface topology [A5, B6, E1, K2]. Sullivan's recent solution of the Fatou-Julia problem shows that this class can also be used very effectively to study problems on the iteration of rational functions [S3, S4].

Higher dimensional quasiconformal mappings were already considered by Lavrentieff in the 1930s [L2]. However, no systematic tool for studying this class was available until 1959 when Loewner introduced the notion of *conformal capacity* to show that $\mathbf{R}^n$ cannot be mapped quasiconformally onto a proper subset of itself [L7].

Subsequently, Gehring, Väisälä, and many others applied Loewner's method and its equivalent extremal length formulation to develop the initial results for quasiconformal mappings in $\mathbf{R}^n$ [G3, V1]. Then in the late 1960s, Reshetnyak and the Finnish school initiated a series of papers which extended the higher dimensional theory to noninjective quasiconformal, or *quasiregular*, mappings [M1, R2, V4], a study which recently resulted in Rickman's remarkable extension of the Picard theorem [R6].

3. *Role played by quasiconformal mappings.* Plane quasiconformal mappings constitute an important tool in complex analysis and they are particularly

valuable in the study of Riemann surfaces and discontinuous groups. Bers's theorem on simultaneous uniformization [B3] is a beautiful application of the measurable Riemann mapping theorem, while Drasin's solution of the inverse problem of Nevanlinna theory [D2] illustrates how this theorem can be used to attack problems of complex analysis in a manner similar to the way the $\bar{\partial}$-equation has been applied in harmonic analysis and several complex variables.

The geometric proofs usually required to establish quasiconformal analogues of results for conformal mappings sometimes yield new insight into classical theorems and methods of complex function theory [L3]. Quasiconformal mappings also arise in exciting and unexpected ways in other parts of mathematics, for example, in harmonic analysis in connection with functions of bounded mean oscillation and singular integrals [B1], and in geometry and elasticity in connection with the injectivity and extension of quasi-isometries.

Higher dimensional quasiconformal mappings offer a new and nontrivial extension of complex analysis to $\mathbf{R}^n$ which is distinct from [N2] and perhaps more geometric and flexible than the analytic theory through several complex variables. These mappings have been applied to solve problems in differential geometry, and they constitute a closed class of mappings, interpolating between homeomorphisms and diffeomorphisms, for which many results of geometric topology hold regardless of dimension. Finally, some of the methods developed to study higher dimensional quasiconformal mappings have found important applications in other branches of mathematics, for example, reverse Hölder inequalities in partial differential equations [G13].

4. *Comments on the above definition.* The quasiconformal mappings studied by Grötzsch and Teichmüller were assumed to be continuously differentiable at all but a finite number of points. Later Ahlfors [A2] and Bers [B2] observed that it was more natural to work with mappings $f\colon D \to D'$ for which one has the important inequality

$$K(f) \leq \liminf_{j \to \infty} K(f_j), \tag{4.1}$$

when $\{f_j\}$ is a sequence of homeomorphisms which converge to $f$ locally uniformly in $D$. Indeed, we defined $K(f)$ as in (1.4), rather than by means of the simpler formula

$$K(f) = \sup_{x \in D} H_f(x), \tag{4.2}$$

just so that (4.1) would hold.

Inequality (4.1) implies that the class of $K$-quasiconformal mappings is closed with respect to locally uniform convergence. Moreover, when $n = 2$, the measurable Riemann mapping theorem implies that every homeomorphism $f$ with $K(f) \leq K$ is the locally uniform limit of continuously differentiable homeomorphisms $f_j$ with $K(f_j) \leq K$ [L3]. When $n = 3$, a quite different argument yields the same conclusion with $K(f_j) \leq \tilde{K}$ where $\tilde{K}$ depends only on $K$ [K1]. The situation when $n > 3$ appears to be open.

If $f: D \to D'$ is a homeomorphism with $K(f) < \infty$, then the Rademacher-Stepanoff theorem and an argument similar to that used by Menchoff imply that $f$ is differentiable with Jacobian $J_f \neq 0$ a.e. in $D$, that $f$ belongs to the Sobolev class $W^n_{1,\mathrm{loc}}(D)$, and that $K(f^{-1}) = K(f)$ [G3]. Thus the inverse of a $K$-quasiconformal mapping is $K$-quasiconformal; similarly, the composition of a $K_1$- and a $K_2$-quasiconformal mapping is $K_1 K_2$-quasiconformal. Though 1-quasiconformal mappings are real analytic, there exists for each $K > 1$ a $K$-quasiconformal self mapping $f$ of $\mathbf{R}^n$ which is not differentiable in a set of Hausdorff dimension $n$.

5. *Remark.* Since there are several excellent expository articles on plane quasiconformal mappings and their connections with Teichmüller spaces [A6, B4, B5, B7], the remainder of this lecture will emphasize the less developed theory in higher dimensions. In Chapter II we consider some basic results and open problems for quasiconformal mappings, comparing what is known for $n = 2$ and for $n > 2$. Then in Chapter III we mention several instances where these mappings arise naturally in other areas of mathematics.

## II. Some results and open problems.

6. *Tools for studying quasiconformal mappings.* A homeomorphism $f: D \to D'$ is quasiconformal if the distortion function $H_f$ in (1.2) is bounded. This is a local restriction and we must find some way to integrate it over $D$ in order to obtain global properties of $f$. In classical complex function theory, this is accomplished by means of the Cauchy integral formula. Though Pompeiu's analogue is sometimes useful in treating plane quasiconformal mappings, the tool most often used to replace the Cauchy formula is the method of extremal length, formulated by Ahlfors and Beurling [A8], and its extension to higher dimensions [F3, V3].

7. *Modulus of a curve family.* Suppose that $\Gamma$ is a family of curves in $\overline{\mathbf{R}}^n$ and let $\mathrm{adm}(\Gamma)$ denote the collection of all Borel measurable functions $\rho: \mathbf{R}^n \to [0, \infty]$ such that $\int_\gamma \rho \, ds \geq 1$ for each locally rectifiable curve $\gamma$ in $\Gamma$. Then

$$\mathrm{mod}(\Gamma) = \inf_{\rho \in \mathrm{adm}(\Gamma)} \int_{\mathbf{R}^n} \rho^n \, dm \quad \text{and} \quad \lambda(\Gamma) = \mathrm{mod}(\Gamma)^{1/(1-n)} \qquad (7.1)$$

are the *conformal modulus* and *extremal length*, respectively, of $\Gamma$.

It is not difficult to see that $\mathrm{mod}(\Gamma)$ is an outer measure on the space of all curve families in $\overline{\mathbf{R}}^n$. Alternatively, if we regard the curves in $\Gamma$ as homogeneous wires, then we may think of $\lambda(\Gamma)$ as the resistance of the family $\Gamma$. In particular, $\mathrm{mod}(\Gamma)$ is large if the curves in $\Gamma$ are short and plentiful, and small otherwise.

The importance of the conformal modulus in the present context is due to its quasi-invariance with respect to quasiconformal mappings.

7.2. THEOREM. *If $f: D \to D'$ is $K$-quasiconformal and if $\Gamma$ is a family of curves which lie in $D$, then*

$$K^{1-n} \mathrm{mod}(\Gamma) \leq \mathrm{mod}(f(\Gamma)) \leq K^{n-1} \mathrm{mod}(\Gamma). \qquad (7.3)$$

Inequality (7.3) plays a key role in the study of quasiconformal mappings. For this reason it is customary to refer to

$$K^*(f) = \max\left(\sup_\Gamma \left(\frac{\mathrm{mod}(f(\Gamma))}{\mathrm{mod}(\Gamma)}\right), \sup_\Gamma \left(\frac{\mathrm{mod}(\Gamma)}{\mathrm{mod}(f(\Gamma))}\right)\right) \qquad (7.4)$$

as the *maximal dilatation* of $f$ and say that $f$ is *$K$-quasiconformal* if $K^*(f) \le K$; here the supremum in (7.4) is taken over all curve families $\Gamma$ in $D$ for which $\mathrm{mod}(\Gamma)$ and $\mathrm{mod}(f(\Gamma))$ are not simultaneously 0 or $\infty$. The inequality

$$K(f)^{n/2} \le K^*(f) \le K(f)^{n-1} \qquad (7.5)$$

shows that this definition yields the same class of quasiconformal mappings and that $K^*(f) = K(f)$ whenever $n = 2$ or $K(f) = 1$.

A homeomorphism $f: \mathbf{R}^n \to \mathbf{R}^n$ is quasiconformal if and only if there exists a constant $c$ such that

$$\limsup_{r \to 0} H_f(x, r) \le c \qquad (7.6)$$

for all $x \in \mathbf{R}^n$. We illustrate the use of (7.3) by establishing a global form of this inequality.

**7.7. THEOREM.** *If $f: \mathbf{R}^n \to \mathbf{R}^n$ is $K$-quasiconformal, then*

$$H_f(x, r) \le c \qquad (7.8)$$

*for all $x \in \mathbf{R}^n$ and $0 < r < \infty$, where $c = c(K, n)$.*

The proof depends on two estimates for the conformal moduli of certain curve families [**G2, G12, V1**].

**7.9. LEMMA.** *If $0 < a < b < \infty$ and if $\Gamma$ is a family of open arcs in $\overline{\mathbf{R}}^n$ which join $\mathbf{S}^{n-1}(0, a)$ to $\mathbf{S}^{n-1}(0, b)$, then*

$$\mathrm{mod}(\Gamma) \le \omega_{n-1}(\log \tfrac{b}{a})^{1-n},$$

*where $\omega_{n-1}$ denotes the $(n-1)$-measure of $\mathbf{S}^{n-1}$.*

**7.10. LEMMA.** *If $C_1$ and $C_2$ are disjoint continua in $\overline{\mathbf{R}}^n$ which join 0 to $\mathbf{S}^{n-1}(0, a)$ and $\infty$ to $\mathbf{S}^{n-1}(0, b)$, respectively, and if $\Gamma$ is the family of all open arcs which join $C_1$ to $C_2$ in $\overline{\mathbf{R}}^n \backslash (C_1 \cup C_2)$, then*

$$\mathrm{mod}(\Gamma) \ge \omega_{n-1}(\log(\lambda_n(\tfrac{b}{a} + 1)))^{1-n},$$

*where $\lambda_n$ depends only on $n$.*

**7.11. COROLLARY.** *If $n > 2$, if $C_1$ and $C_2$ are disjoint, linked continua in $\overline{\mathbf{R}}^n$ and if $\Gamma$ is the family of all open arcs which join $C_1$ and $C_2$ in $\overline{\mathbf{R}}^n \backslash (C_1 \cup C_2)$, then $\mathrm{mod}(\Gamma) \ge c$ where $c = c(n) > 0$.*

PROOF OF THEOREM 7.7. By performing preliminary translations, we may assume that $x = 0$ and $f(0) = 0$. Let $m$ and $M$ denote the minimum and maximum values assumed by $|f|$ on $\mathbf{S}^{n-1}(0, r)$ and suppose that $m < M$. Next set

$$C_1 = \{x \in \mathbf{R}^n: |f(x)| \le m\}, \qquad C_2 = \{x \in \mathbf{R}^n: |f(x)| \ge M\} \cup \{\infty\},$$

and let $\Gamma$ be the family of open arcs which join $C_1$ and $C_2$ in $\overline{\mathbf{R}}^n \backslash (C_1 \cup C_2)$. Then the above estimates and (7.3) imply that

$$\omega_{n-1}(\log 2\lambda_n)^{1-n} \le \text{mod}(\Gamma) \le K^{n-1} \text{mod}(f(\Gamma)) \le K^{n-1}\omega_{n-1}(\log(M/m))^{1-n}$$

and we obtain (7.8) with $c = (2\lambda_n)^K$.

**8. Mapping problems.** A basic question in this area is to decide when two domains in $\overline{\mathbf{R}}^n$ are quasiconformally equivalent, i.e., if one can be mapped quasiconformally onto the other. Since the general case is quite difficult even when $n = 2$, we consider here the simpler problem of characterizing the domains $D$ in $\overline{\mathbf{R}}^n$ which are quasiconformally equivalent to the unit ball $\mathbf{B}^n$. The Riemann mapping theorem and the estimates in Lemmas 7.9 and 7.10 yield a complete answer when $n = 2$.

**8.1. THEOREM.** *A domain $D$ in $\overline{\mathbf{R}}^2 <$ is quasiconformally equivalent to $\mathbf{B}^2$ if and only if $\partial D$ is a nondegenerate continuum.*

No such characterization exists in higher dimensions. Indeed, the domains $D_3$ and $D_4$ in (8.6) below show that when $n > 2$, there is no way to decide whether the image of $\mathbf{B}^n$ under a self homeomorphism of $\overline{\mathbf{R}}^n$ is quasiconformally equivalent to $\mathbf{B}^n$ by looking only at its boundary.

The following sufficient condition is a consequence of methods used to treat the higher dimensional Schoenflies problem [**G5, M2**].

**8.2. THEOREM.** *A domain $D$ in $\overline{\mathbf{R}}^n$ is quasiconformally equivalent to $\mathbf{B}^n$ if there exist closed sets $E \subset D$, $E' \subset \mathbf{B}^n$ and a quasiconformal mapping $g: D \backslash E \to \mathbf{B}^n \backslash E'$ such that $|g(x)| \to 1$ as $x \to \partial D$ in $D$.*

As in the topological case, localized versions of Theorem 8.2 can be established when $D$ is a Jordan domain in $\overline{\mathbf{R}}^n$, i.e., when $\partial D$ is homeomorphic to $\mathbf{S}^{n-1}$ [**B10, G1**].

**8.3. COROLLARY.** *If $D$ is a domain in $\mathbf{R}^n$ and if $D$ is diffeomorphic to $\mathbf{S}^{n-1}$, then $D$ is quasiconformally equivalent to $\mathbf{B}^n$.*

It is easy to construct a domain in $\mathbf{R}^n$ which is quasiconformally equivalent to $\mathbf{B}^n$ and does not have a tangent plane at any point of its boundary [**G12**]. Thus the sufficient condition in Corollary 8.3 is far from necessary.

A necessary condition for quasiconformal equivalence to $\mathbf{B}^n$ depends on the following refinement of the notion of local connectivity. A set $E \subset \overline{\mathbf{R}}^n$ is said to be *linearly locally connected* if there exists a constant $c$, $1 \le c < \infty$, such that for each $x \in \mathbf{R}^n$ and $0 < r < \infty$

$$\begin{aligned} E \cap \overline{\mathbf{B}}^n(x,r) &\quad \text{lies in a component of } E \cap \overline{\mathbf{B}}^n(x,cr), \\ E \backslash \mathbf{B}^n(x,r) &\quad \text{lies in a component of } E \backslash \mathbf{B}^n(x,r/c). \end{aligned} \qquad (8.4)$$

Then an argument based again on inequality (7.3) and the estimates in Lemma 7.9 and Corollary 7.11 implies the following result [**G7, G12**].

**8.5. THEOREM.** *If $n > 2$ and if $D$ in $\overline{\mathbf{R}}^n$ is quasiconformally equivalent to $\mathbf{B}^n$, then $\overline{\mathbf{R}}^n \backslash D$ is linearly locally connected.*

Theorem 8.5 yields many simple domains in $\mathbf{R}^n$ which are homeomorphic, but not quasiconformally equivalent, to $\mathbf{B}^n$. For example, let

$$
\begin{aligned}
D_1 &= \{x \in \mathbf{R}^n : r < 1, |x_n| < \infty\}, \\
D_2 &= \{x \in \mathbf{R}^n : r < \infty, |x_n| < 1\}, \\
D_3 &= \{x \in \mathbf{R}^n : x_n > \min(r^{1/2}, 1)\}, \\
D_4 &= \{x \in \mathbf{R}^n : x_n < \min(r^{1/2}, 1)\},
\end{aligned}
\tag{8.6}
$$

where $x = (x_1, \ldots, x_n)$ and $r = (x_1^2 + \cdots + x_{n-1}^2)^{1/2}$. Then explicit constructions yield homeomorphisms which map $D_1$ and $D_3$ quasiconformally onto $\mathbf{B}^n$. On the other hand when $n > 2$, $\overline{\mathbf{R}}^n \backslash D_2$ and $\overline{\mathbf{R}}^n \backslash D_4$ are not linearly locally connected and hence $D_2$ and $D_4$ are not quasiconformally equivalent to $\mathbf{B}^n$.

The necessary condition in Theorem 8.5 is not sufficient and the problem of finding sharp geometric criteria for testing quasiconformal equivalence to $\mathbf{B}^n$ remains a most interesting open question.

9. *Homeomorphic and quasiconformal extensions.* Suppose that $D$ and $D'$ are domains in $\overline{\mathbf{R}}^n$ and that $f : D \to D'$ is quasiconformal. We consider next under what circumstances $f$ admits a homeomorphic extension to $\overline{D}$ or a quasiconformal extension to $\overline{\mathbf{R}}^n$.

**9.1. THEOREM.** *If $D$ and $D'$ are simply-connected domains of hyperbolic type in $\overline{\mathbf{R}}^2$, then each quasiconformal $f : D \to D'$ has a homeomorphic extension to $\overline{D}$ if and only if $D$ and $D'$ are Jordan domains.*

The sufficiency in Theorem 9.1 follows from a theorem of Ahlfors [A2] and the necessity from [E3]. In higher dimensions we have the following result [V2].

**9.2. THEOREM.** *If $D$ and $D'$ are Jordan domains in $\overline{\mathbf{R}}^n$ and if $D$ is quasiconformally equivalent to $\mathbf{B}^n$, then each quasiconformal $f : D \to D'$ has a homeomorphic extension to $\overline{D}$.*

When $n = 2$, the second hypothesis in Theorem 9.2 is superfluous since every Jordan domain is conformally equivalent to $\mathbf{B}^2$. When $n > 2$, this is not the case as seen by the examples in (8.6), and Theorem 9.2 does not hold without this additional restriction [K3].

As to the problem of quasiconformal extension to $\overline{\mathbf{R}}^n$, we say that a set $E$ in $\overline{\mathbf{R}}^2$ is a *$K$-quasidisk* or *$K$-quasicircle* if it is the image of $\mathbf{B}^2$ or $\mathbf{S}^1$, respectively, under a $K$-quasiconformal self mapping of $\overline{\mathbf{R}}^2$. By a theorem of Ahlfors, a Jordan domain $D$ is a quasidisk if and only if there exists a constant $c$ such that

$$
\min_{j=1,2} \operatorname{dia}(\gamma_j) \le c |z_1 - z_2|
\tag{9.3}
$$

for each $z_1, z_2 \in \partial D$, where $\gamma_1$ and $\gamma_2$ denote the components of $\partial D \backslash \{z_1, z_2\}$ [A3].

**9.4. THEOREM.** *If $D$ and $D'$ are Jordan domains in $\overline{\mathbf{R}}^2$, then each quasiconformal $f: D \to D'$ has a quasiconformal extension to $\overline{\mathbf{R}}^2$ if and only if $D$ and $D'$ are quasidisks.*

A simply-connected domain $D$ in $\overline{\mathbf{R}}^2$ is a quasidisk if and only if it is linearly locally connected. Hence this notion also arises in connection with quasiconformal extension.

The sufficiency in Theorem 9.4 is due to Ahlfors [A2] and the necessity to Rickman [R5]. A higher dimensional analogue of this result is as follows [G4, V5].

**9.5. THEOREM.** *If $n > 2$ and if $D$ is a Jordan domain in $\overline{\mathbf{R}}^n$, then each quasiconformal $f: D \to \mathbf{B}^n$ has a quasiconformal extension to $\overline{\mathbf{R}}^n$ if and only if $D^* = \overline{\mathbf{R}}^n \backslash \overline{D}$ is quasiconformally equivalent to $\mathbf{B}^n$.*

Thus the problem of quasiconformal extension in higher dimensions differs from the plane case in two respects. First, when $n = 2$, the exterior $D^*$ of every Jordan domain $D$ is quasiconformally equivalent to $\mathbf{B}^2$; this is not true when $n > 2$ as we observed above. Second, when $n > 2$, each quasiconformal $f: D \to \mathbf{B}^n$ has a quasiconformal extension to $\overline{\mathbf{R}}^n$ whenever $D^*$ is quasiconformally equivalent to $\mathbf{B}^n$; this is not true when $n = 2$ since there exist Jordan domains $D$ which do not satisfy condition (9.3) and hence are not quasidisks.

10. *Boundary correspondence.* We turn to the problem of characterizing the boundary mappings induced by quasiconformal self mappings of balls and halfspaces. For $n \geq 2$ let $\mathbf{H}^n$ denote the upper halfspace $\{x \in \mathbf{R}^n: x_n > 0\}$. Then each quasiconformal $f: \mathbf{H}^n \to \mathbf{H}^n$ has a quasiconformal extension $\tilde{f}$ to $\overline{\mathbf{R}}^n$ whose restriction to $\partial \mathbf{H}^n$ is a self homeomorphism $\varphi$ of $\overline{\mathbf{R}}^{n-1}$. The problem of studying such boundary correspondences was initiated by Beurling and Ahlfors [B8].

**10.1. THEOREM.** *A homeomorphism $\varphi: \overline{\mathbf{R}}^1 \to \overline{\mathbf{R}}^1$ with $\varphi(\infty) = \infty$ is the boundary correspondence for a quasiconformal self mapping $f$ of $\mathbf{H}^2$ with $\tilde{f}(\infty) = \infty$ if and only if there exists a constant $c$ such that*

$$\frac{1}{c} \leq \frac{\varphi(x+r) - \varphi(x)}{\varphi(x) - \varphi(x-r)} \leq c \tag{10.2}$$

*for all $x \in \mathbf{R}^1$ and $0 < r < \infty$.*

Inequality (10.2) is equivalent to the requirement that $H_\varphi(x,r) \leq c$. This condition is replaced by its local form $H_\varphi(x) \leq c$, or that $\varphi$ is quasiconformal, in the higher dimensional analogue of Theorem 10.1.

**10.3. THEOREM.** *When $n > 2$, a homeomorphism $\varphi: \overline{\mathbf{R}}^{n-1} \to \overline{\mathbf{R}}^{n-1}$ is the boundary correspondence of a quasiconformal self mapping $f$ of $\mathbf{H}^n$ if and only if $\varphi$ is itself quasiconformal.*

The necessity in Theorems 10.1 and 10.3 follows, respectively, from inequalities (7.8) and (7.6) and the fact that

$$H_\varphi(x,r) \leq H_{\tilde{f}}(x,r) \quad \text{and} \quad H_\varphi(x) \leq H_{\tilde{f}}(x) \tag{10.4}$$

for relevant $x$ and $r$.

Beurling and Ahlfors established the sufficiency in Theorem 10.1 by showing that the formula

$$f(x_1, x_2) = \frac{1}{2x_2} \int_0^{x_2} \left( \varphi(x_1 + t) + \varphi(x_1 - t) \right) dt$$
$$+ \frac{i}{2x_2} \int_0^{x_2} \left( \varphi(x_1 + t) - \varphi(x_1 - t) \right) dt \tag{10.5}$$

defines a quasiconformal extension of $\varphi$ to $\mathbf{H}^2$.

Ahlfors [A4] modified this construction and used the fact that each quasiconformal $\varphi \colon \overline{\mathbf{R}}^2 \to \overline{\mathbf{R}}^2$ can be written as the composition of mappings with small dilatation (see Corollary 11.4) to obtain a quasiconformal extension of $\varphi$ to $\mathbf{H}^3$ and thus prove the sufficiency in Theorem 10.3 when $n = 3$. Next Carleson [C1] employed quite different methods from three-dimensional topology to extend each quasiconformal $\varphi \colon \overline{\mathbf{R}}^3 \to \overline{\mathbf{R}}^3$ to $\mathbf{H}^4$. Finally, Tukia and Väisälä [T5] started from an idea of Carleson's and applied results of Sullivan's [S1] to establish the sufficiency in Theorem 10.3 for general $n$.

After composition with suitable Möbius transformations, (10.2) yields a cross ratio characterization for the boundary mappings $\varphi \colon \partial D \to \partial D$ induced by arbitrary quasiconformal self mappings of a disk or halfplane $D$ in $\mathbf{R}^2$, and (10.5) gives an explicit quasiconformal extension $T(\varphi) \colon D \to D$ of each such correspondence $\varphi$. Tukia [T4] recently settled an important problem in Teichmüller theory by showing that if $G$ is a subgroup of Möb$(D)$, the group of all Möbius self mappings of $D$, then each $G$-compatible boundary correspondence $\varphi \colon \partial D \to \partial D$ has a $G$-compatible quasiconformal extension to $D$. Douady and Earle [D1] extended this work by exhibiting a *conformally natural* quasiconformal extension operator $T_0$ such that

$$g \circ T_0(\varphi) \circ h = T_0(g \circ \varphi \circ h) \tag{10.6}$$

for each homeomorphism $\varphi \colon \partial D \to \partial D$ and all $g, h \in$ Möb$(D)$. This beautiful operator should yield many new results in the area; see [E2].

If $D$ is a ball or halfspace in $\mathbf{R}^n$ where $n > 2$, then the method of Douady and Earle assigns to each homeomorphism $\varphi \colon \partial D \to \partial D$ a continuous extension $T_0(\varphi) \colon D \to D$ for which (10.6) holds. However, $T_0(\varphi)$ will, in general, be neither quasiconformal nor injective except when $K(\varphi)$ is small, i.e., $K(\varphi) \leq K_n$ where $K_n$ depends only on $n$. It would be interesting to know if every quasiconformal $\varphi$ has a conformally natural quasiconformal extension.

**11. Measurable Riemann mapping theorem and decomposition.** If $f \colon D \to D'$ is quasiconformal, then $f$ has a nonsingular differential $df \colon \mathbf{R}^n \to \mathbf{R}^n$ at almost all $x \in D$. At each such $x$, $df = df(x)$ maps an ellipsoid $E_f = E_f(x)$ about 0 with minimum axis length 1 onto an $(n-1)$-sphere about 0. Then $H_f(x)$ is the maximum axis length of $E_f(x)$ and the maximum stretching under $f$ at $x$ occurs in the directions of the smallest axes of $E_f(x)$. If $g \colon D' \to D''$ is quasiconformal, then $g$ is conformal if and only if $E_{g \circ f} = E_f$ a.e. in $D$ by Theorem 1.5, and $E_f$ determines $f$ up to postcomposition with a conformal mapping.

When $n = 2$ and $f$ is sense preserving, $E_f$ is determined by the *Beltrami coefficient* or *complex dilatation*

$$\mu_f(x) = f_{\bar{z}}/f_z, \qquad x = x_1 + ix_2, \tag{11.1}$$

of $f$ at $x$. In particular, $\mu_f$ is measurable with

$$|\mu_f(x)| = \frac{H_f(x) - 1}{H_f(x) + 1}, \qquad \|\mu_f\|_{L^\infty} = \frac{K(f) - 1}{K(f) + 1} < 1, \tag{11.2}$$

and $\mu_{g \circ f} = \mu_f$ a.e. in $D$ if and only if $g: D' \to D''$ is conformal. Moreover, in dimension two it is possible to prescribe the dilatation $\mu_f$, and hence the ellipse $E_f$, at almost every $x \in D$ [**A7**].

11.3. MEASURABLE RIEMANN MAPPING THEOREM. *If $\mu$ is measurable with $\|\mu\|_{L^\infty} < 1$ in $\bar{\mathbf{R}}^2$, then there exists a quasiconformal self mapping $f = f_\mu$ of $\bar{\mathbf{R}}^2$ with $\mu_f = \mu$ a.e. If $f$ is normalized to fix three points, then $f$ is unique and depends holomorphically on $\mu$.*

Theorem 11.3 is of fundamental importance in studying the complex structure on Teichmüller space. It can also be a powerful tool for attacking other problems of complex analysis. One example is the solution of the inverse problem of Nevanlinna theory [**D2**] where Drasin first constructed a locally quasiconformal function $g$ with prescribed defects, and then applied Theorem 11.3 to obtain a quasiconformal self mapping $f$ of $\mathbf{R}^2$ so that $g \circ f$ was meromorphic with the same defects as $g$. A second example is Sullivan's recent solution [**S3**] of the Fatou-Julia problem on wandering domains where Theorem 11.3 was used to construct a large real analytic family of quasiconformal deformations of a given rational function.

The following is an important consequence of Theorem 11.3.

11.4. COROLLARY. *If $n = 2$ and $\varepsilon > 0$, then each quasiconformal $f: D \to D'$ can be written in the form $f = f_1 \circ \cdots \circ f_m$ where $K(f_j) < 1 + \varepsilon$ for $j = 1, \ldots, m$ and $m = m(\varepsilon, K(f))$.*

There is no analogue of Theorem 11.3 in higher dimensions. Moreover, when $n > 2$, examples show that Corollary 11.4 is almost certainly not true without further restrictions on the domain $D$. It is an important open problem to decide if some higher dimensional form of this result holds, even for the case where $D = D' = \bar{\mathbf{R}}^n$.

12. *Quasiconformal groups.* A group $G$ of self homeomorphisms of $\bar{\mathbf{R}}^n$ is said to be *discrete* if $G$ contains no sequence of elements which converge to the identity uniformly in $\bar{\mathbf{R}}^n$, and *$K$-quasiconformal* if $K(g) \leq K$ for each $g$ in $G$. Though the family of quasiconformal groups contains all Möbius groups, Theorem 11.3 can be used to show that this larger family does not exhibit new phenomena when $n = 2$ [**S2**, **T2**].

12.1. THEOREM. *When $n = 2$, each quasiconformal group $G$ can be written in the form $G = f^{-1} \circ H \circ f$, where $H$ is a Möbius group and $f$ a quasiconformal self mapping of $\bar{\mathbf{R}}^2$.*

The situation is different in higher dimensions, and for each $n > 2$ there exists a quasiconformal group which is not even isomorphic as a topological group to a Möbius group [**T3**]. Nevertheless, the following convergence property allows one to establish quasiconformal analogues of many basic properties of Möbius groups [**G10**].

**12.2. THEOREM.** *If $G$ is a discrete quasiconformal group, then for each infinite sequence of distinct elements in $G$ there exists a subsequence $\{g_j\}$ and points $x_0, y_0$ in $\overline{\mathbf{R}}^n$ such that $g_j \to y_0$ locally uniformly in $\overline{\mathbf{R}}^n \backslash \{x_0\}$ and $g_j^{-1} \to x_0$ locally uniformly in $\overline{\mathbf{R}}^n \backslash \{y_0\}$.*

Suppose that $G$ is a group of self homeomorphisms of $\overline{\mathbf{R}}^n$. We say that $G$ is a *discrete convergence* group if it satisfies the conclusion of Theorem 12.2, and that an element $g$ of $G$ is *elliptic* if it is of finite order or periodic, and *parabolic* or *loxodromic* if it has infinite order and one or two fixed points, respectively. The *limit set* $L(G)$ is the complement of the ordinary set $O(G)$, the set of $x \in \overline{\mathbf{R}}^n$ which have a neighborhood $U$ such that $g(U) \cap U \neq \varnothing$ for at most finitely many $g \in G$. Finally, $G$ is *properly discontinuous* in an open set $O$ if for each compact $F \subset O$, $g(F) \cap F \neq \varnothing$ for at most finitely many $g \in G$ [**G10**].

**12.3. THEOREM.** *Suppose that $G$ is a discrete convergence group. Then each element of $G$ is elliptic, parabolic, or loxodromic, and the limit set $L(G)$ is nowhere dense or equal to $\overline{\mathbf{R}}^n$. Moreover if $\operatorname{card}(L(G)) > 2$, then $L(G)$ is perfect, $L(G)$ lies in the closure of each nonempty $G$-invariant set, and the set of fixed point pairs of loxodromic elements in $G$ is dense in $L(G) \times L(G)$.*

Though discrete convergence groups resemble Möbius groups in many respects, examples exist which show that they need not be topologically conjugate to Möbius groups [**F2, G10**]. They also occur quite naturally in situations which have nothing to do with Möbius or quasiconformal groups.

**12.4. THEOREM.** *A group $G$ of self homeomorphisms of $\overline{\mathbf{R}}^n$ is a discrete convergence group if it is properly discontinuous in $\overline{\mathbf{R}}^n \backslash E$, where $E$ is closed and totally disconnected.*

It will be interesting to see how much of the classical theory of kleinian groups carries over for this general class of groups.

**13. *Hölder continuity and integrability.*** Theorem 12.2 can be deduced from (4.1) and the following estimate for change in the chordal distance $q$ in (1.1) under a quasiconformal mapping [**G7**].

**13.1. THEOREM.** *If $f: D \to D'$ is $K$-quasiconformal and if $\overline{\mathbf{R}}^n \backslash D \neq \varnothing$, then*

$$q(f(x), f(y)) q(\overline{\mathbf{R}}^n \backslash D') \leq c(q(x,y)/q(x, \partial D))^{1/K} \qquad (13.2)$$

*for $x, y$ in $D$, where $q(E)$ denotes the chordal diameter of $E$ and $c = c(n)$.*

Theorem 13.1 is a consequence of (7.3) and Lemmas 7.9 and 7.10. When $D, D' \subset \mathbf{R}^n$, it implies that each $K$-quasiconformal $f: D \to D'$ is locally Hölder continuous with exponent $1/K$ and hence that these mappings interpolate between diffeomorphisms and homeomorphisms for $1 \leq K < \infty$. This fact is also reflected in the integrability of the Jacobian $J_f$ of $f$.

13.3. THEOREM. *If* $f: D \to D'$ *is* $K$-*quasiconformal where* $D, D' \subset \mathbf{R}^n$, *then* $J_f$ *is locally* $L^p$-*integrable in* $D$ *for* $1 \leq p < p(K, n)$, *where*

$$p(K, n) \leq K/(K - 1), \qquad \lim_{K \to 1} p(K, n) = \infty. \tag{13.4}$$

Bojarski [B9] established the existence of the exponent $p(K, n)$ for $n = 2$ by applying the Calderón-Zygmund inequality to the Beurling transform in (14.7). The proof for $n > 2$ was based on the fact that $g = |J_f|$ satisfies the reverse Hölder inequality

$$\frac{1}{m(Q)} \int_Q g \, dm \leq c \left( \frac{1}{m(Q)} \int_Q g^{1/n} \, dm \right)^n, \qquad c = c(K, n), \tag{13.5}$$

for each $n$-cube $Q$ in $D$ with $\mathrm{dia}(f(Q)) < d(f(Q), \partial D')$, and on a lemma to the effect that if (13.5) holds for all $n$-cubes $Q$ contained in an $n$-cube $Q'$, then $g$ belongs to $L^p(Q')$ for $1 \leq p < p(c, n)$ [G6]. These results were sharpened in [I2, R4] to obtain the second part of (13.4); the example

$$f(x) = |x|^{-a} x, \qquad a = (K - 1)/K, \tag{13.6}$$

gives the first part of (13.4). There is reason to suspect that one can take $p(K, n) = K/(K - 1)$ in Theorem 13.3. However, this has not been established even for the case $n = 2$.

## III. Connections with other areas of mathematics.

14. *Harmonic and functional analysis.* Quasiconformal mappings are encountered in harmonic analysis through their connections with functions of bounded mean oscillation and singular integrals.

A function $u$ is said to be of *bounded mean oscillation* in a domain $D \subset \mathbf{R}^n$, or in BMO($D$), if $u$ is locally integrable and

$$\|u\|_{\mathrm{BMO}(D)} = \sup_B \frac{1}{m(B)} \int_B |u - u_B| \, dm < \infty, \tag{14.1}$$

where the supremum is taken over all $n$-balls $B$ with $\overline{B} \subset D$ and

$$u_B = \frac{1}{m(B)} \int_B u \, dm. \tag{14.2}$$

The class BMO was introduced by John and Nirenberg [J3] in connection with John's work in elasticity [J1], and it gained great prominence when Fefferman showed that BMO($\mathbf{R}^n$) is the dual of the Hardy space $H^1(\mathbf{R}^n)$ [F1].

The following relations between the class BMO and quasiconformal mappings are due to Reimann [R1].

**14.3. THEOREM.** *If $f: D \to D'$ is $K$-quasiconformal where $D, D' \subset \mathbf{R}^n$, then $\| \log J_f \|_{\mathrm{BMO}(D)} \leq c$ where $c = c(K, n)$.*

**14.4. THEOREM.** *Suppose that $f: D \to D'$ is a homeomorphism where $D, D' \subset \mathbf{R}^n$. Then $f$ is quasiconformal if and only if there exists a constant $c$ such that*

$$\frac{1}{c}\|u\|_{\mathrm{BMO}(G')} \leq \|u \circ f\|_{\mathrm{BMO}(G)} \leq c\|u\|_{\mathrm{BMO}(G')} \tag{14.5}$$

*for each subdomain $G$ of $D$ and each $u$ continuous in $G' = f(G)$.*

Theorem 14.3 and the necessity in Theorem 14.4 follow from the fact that $g = |J_f|$ satisfies the reverse Hölder inequality in (13.5). The sufficiency in Theorem 14.4 is a variant, due to Astala [A9], of Reimann's original result.

Theorem 14.4 characterizes quasiconformal mappings as the homeomorphisms which preserve the class BMO. The following result [J4] characterizes quasidisks in terms of extension properties for BMO.

**14.6. THEOREM.** *If $D$ is a simply-connected domain of hyperbolic type in $\mathbf{R}^2$, then each function $u$ in $\mathrm{BMO}(D)$ has a BMO extension to $\mathbf{R}^2$ if and only if $D$ is a quasidisk.*

Next the best possible exponents for Jacobian integrability and area distortion for plane quasiconformal mappings are closely connected with sharp constants in two inequalities for the Beurling transform

$$Tg(x) = -\frac{1}{\pi} \int_{\mathbf{R}^2} \frac{g(y)}{(x-y)^2}\, dm. \tag{14.7}$$

For example, $T$ is a bounded operator on $L^p(\mathbf{R}^2)$ with

$$\|T\|_p = \sup_g \frac{\|Tg\|_{L^p(\mathbf{R}^2)}}{\|g\|_{L^p(\mathbf{R}^2)}} \geq \max\left(p-1, \frac{1}{p-1}\right) \tag{14.8}$$

for $1 < p < \infty$ and $\|T\|_2 = 1$, and there is reason to believe that

$$\liminf_{p \to \infty} \frac{1}{p}\|T\|_p = 1. \tag{14.9}$$

If true, this would yield the sharp upper bound $p(K, 2) = K/(K-1)$ for the integrability of the Jacobian of a plane quasiconformal mapping discussed in §13 [I1].

Next, one can show that there exist constants $a$ and $b$ such that

$$\int_{\mathbf{B}^2} |T\chi_E(x)|\, dm \leq am(E)\log(\pi/m(E)) + bm(E) \tag{14.10}$$

for each measurable $E \subset \mathbf{B}^2$. This inequality can be combined with Theorem 11.3 to prove that

$$\frac{m(f(E))}{\pi} \leq c \left(\frac{m(E)}{\pi}\right)^{K^{-a}}, \tag{14.11}$$

$c = c(K) = 1 + O(K-1)$ as $K \to 1$, for each $K$-quasiconformal $f: \mathbf{B}^2 \to \mathbf{B}^2$ with $f(0) = 0$ and each measurable set $E \subset \mathbf{B}^2$ [G11]. Moreover, the above reasoning can be reversed to show that if (14.11) holds for a given constant $a$, then so does (14.10). It is conjectured that both hold with $a = 1$. If so, this would again imply that $p(K, 2) = K/(K - 1)$.

Finally, the problem of quasiconformal equivalence of domains can be reformulated in terms of function algebras. Given a domain $D$ in $\mathbf{R}^n$, we let $A(D)$ denote the algebra of functions $u \in C(D) \cap W_n^1(D)$ with norm

$$\|u\| = \|u\|_{L^\infty(D)} + \|\nabla u\|_{L^n(D)}, \qquad (14.12)$$

the so-called *Royden algebra* of $D$. We then have the following result [L5, L6].

14.13. THEOREM. *Two domains $D$ and $D'$ in $\mathbf{R}^n$ are quasiconformally equivalent if and only if $A(D)$ and $A(D')$ are isomorphic as algebras.*

Little is known about the structure of these algebras and it may be that geometric methods used to determine quasiconformal equivalence will yield more information about them than vice versa.

15. *Quasi-isometries and elasticity.* A mapping $f\colon E \subset \mathbf{R}^n \to \mathbf{R}^n$ is an $L$-*quasi-isometry* in $E$ if

$$\frac{1}{L}|x_1 - x_2| \leq |f(x_1) - f(x_2)| \leq L|x_1 - x_2| \qquad (15.1)$$

for $x_1, x_2 \in E$; $f$ is a *local $L$-quasi-isometry* in $E$ if for each $L' > L$, each $x \in E$ has a neighborhood $U$ such that $f$ is an $L'$-quasi-isometry in $E \cap U$.

If $f$ is quasi-isometric in a domain $D$, then $f$ is quasiconformal by (1.2) and (1.4); the mapping in (13.6) shows that the converse is false. Nevertheless, quasiconformal homeomorphisms arise in questions concerning extension and injectivity of these mappings.

15.2. THEOREM. *If $n \neq 4$, then a quasi-isometry $f$ of $E$ has a quasi-isometric extension to $\mathbf{R}^n$ if and only if $f$ has a quasiconformal extension to $\mathbf{R}^n$.*

Theorem 15.2 [T6] gives a criterion for extension in terms of the mapping $f$. There is also a criterion in terms of the set $E$ when $E$ is a Jordan curve [G9].

15.3. THEOREM. *If $C$ is a Jordan curve in $\mathbf{R}^2$, then each quasi-isometry $f$ of $C$ has a quasi-isometric extension to $\mathbf{R}^2$ if and only if $C$ is a quasicircle.*

For each domain $D \subset \mathbf{R}^n$ let $L(D)$ denote the supremum of the numbers $L \geq 1$ with the property that each local $L$-quasi-isometry $f$ in $D$ is injective there. The constant $L(D)$ has a physical interpretation if we think of $D$ as an elastic body and $f$ as the deformation experienced by $D$ when subjected to a force field. Requiring that $f$ be a local $L$-quasi-isometry bounds the strain in $D$ under the force field and $L(D)$ measures the critical strain in $D$ before $D$ collapses onto itself.

Little is known about this constant except that $2^{1/4} \leq L(D) \leq 2^{1/2}$ whenever $D$ is a ball or halfspace [J2]. However, we can characterize a large class of plane domains for which $L(D) > 1$ [G8].

15.4. THEOREM. *If $D$ is a simply-connected proper subdomain of $\mathbf{R}^2$, then $L(D) > 1$ if and only if $D$ is a quasidisk.*

15.5. COROLLARY. *If $f$ is a local $L$-quasi-isometry of a bounded simply-connected domain in $\mathbf{R}^2$ and if $L < L(D)$, then $f$ has an $M$-quasi-isometric extension to $\mathbf{R}^2$ where $M = M(L, L(D))$.*

Corollary 15.5 says that the shape of a deformed simply-connected plane elastic body is roughly the same as that of the original provided the strain does not attain the critical value. It would be interesting to obtain a higher dimensional analogue of this result.

16. *Complex analysis.* Quasiconformal mappings sometimes arise in function-theoretic problems which appear to be completely unrelated to this class. An excellent example is Teichmüller's theorem [T1] which relates the extremal quasiconformal mappings between two Riemann surfaces with the quadratic differentials on these surfaces.

For a more elementary example, suppose that $f$ is meromorphic in a simply-connected domain $D$ of hyperbolic type in $\overline{R}^2$ and let

$$S_f = \left( \frac{f''}{f'} \right)' - \frac{1}{2} \left( \frac{f''}{f'} \right)^2. \tag{16.1}$$

By a theorem of Nehari [N1], $f$ is injective whenever $D$ is a disk or halfplane and $|S_f| \le 2\rho_D^2$ in $D$. Here $\rho_D$ is the hyperbolic metric in $D$ given by

$$\rho_D(z) = |g'(z)|(1 - |g(z)|^2)^{-1} \tag{16.2}$$

where $g \colon D \to \mathbf{B}^2$ is conformal. It is natural to ask: for which other domains $D$ does such a result hold? That is, for which $D$ is $\sigma(D) > 0$, where $\sigma(D)$ denotes the supremum of the numbers $a \ge 0$ such that $f$ is injective whenever $f$ is meromorphic with $|S_f| \le a\rho_D^2$ in $D$?

The answer involves quasiconformal mappings and yields a new characterization of Bers's universal Teichmüller space [B5].

16.3. THEOREM. *$\sigma(D) > 0$ if and only if $D$ is a quasidisk.*

17. *Differential geometry and topology.* Some of the results mentioned in Chapter II have important applications in differential geometry. For example, Theorem 1.5 and the necessity in Theorem 10.3 are key steps in the original proof of Mostow's rigidity theorem [M5].

17.1. THEOREM. *If $n > 2$ and if $M$ and $M'$ are diffeomorphic compact Riemannian $n$-manifolds with constant negative curvature, then $M$ and $M'$ are conformally equivalent.*

Similarly, the equicontinuity property for quasiconformal mappings implied by Theorem 13.1 is an important tool in establishing the following conjecture of Lichnerowicz [L4].

17.2. THEOREM. *If $n \geq 2$ and if $M$ is a compact Riemannian $n$-manifold not conformally equivalent to a sphere, then the group $C(M)$ of conformal self mappings of $M$ is compact in the topology of uniform convergence.*

The work of Earle and Eells [E1] on the diffeomorphism group of a surface and Bers's proof [B6] of Thurston's theorem on the classification of self mappings of surfaces illustrate how quasiconformal mappings can be applied to problems in surface topology. Sullivan showed [S1] that the Schoenflies theorem, the annulus conjecture and the component problem hold for quasiconformal mappings in all dimensions. The results of this fundamental paper suggest that quasiconformal mappings may prove to be an important, intermediate category of maps between homeomorphisms and diffeomorphisms.

## REFERENCES

[A1] L. V. Ahlfors, *Zur theorie der Überlagerungsflächen*, Acta Math. **65** (1935), 157–194.

[A2] ____, *On quasiconformal mappings*, J. Analyse Math. **3** (1953/54), 1–58.

[A3] ____, *Quasiconformal reflections*, Acta Math. **109** (1963), 291–301.

[A4] ____, *Extension of quasiconformal mappings from two to three dimensions*, Proc. Nat. Acad. Sci. U.S.A. **51** (1964), 768–771.

[A5] ____, *Finitely generated Kleinian groups*, Amer. J. Math. **86** (1964), 413–429.

[A6] ____, *Quasiconformal mappings, Teichmüller spaces, and Kleinian groups*, Proc. Internat. Congr. Math. (Helsinki, 1978), Acad. Sci. Fennica, Helsinki, 1980, pp. 71–84.

[A7] L. V. Ahlfors and L. Bers, *Riemann's mapping theorem for variable metrics*, Ann. of Math. **72** (1960), 385–404.

[A8] L. V. Ahlfors and A. Beurling, *Conformal invariants and function-theoretic null sets*, Acta Math. **83** (1950), 101–129.

[A9] K. Astala, *A remark on quasi-conformal mappings and BMO-functions*, Michigan Math. J. **30** (1983), 209–212.

[B1] A. Baernstein II and J. J. Manfredi, *Topics in quasiconformal mapping*, Topics in Modern Harmonic Analysis, Istituto Nazionale di Alta Mathematica, Roma, 1983, pp. 819–862.

[B2] L. Bers, *Quasiconformal mappings and Teichmüller's theorem*, Analytic Functions, Princeton Univ. Press, Princeton, N.J., 1960, pp. 89–119.

[B3] ____, *Uniformization by Beltrami equations*, Comm. Pure Appl. Math. 14 (1961), 215–228.

[B4] ____, *Uniformization, moduli, and Kleinian groups*, Bull. London Math. Soc. **4** (1972), 257–300.

[B5] ____, *Quasiconformal mappings, with applications to differential equations, function theory and topology*, Bull. Amer. Math. Soc. **83** (1977), 1083–1100.

[B6] ____, *An extremal problem for quasiconformal mappings and a problem of Thurston*, Acta Math. **141** (1978), 73–98.

[B7] ____, *Finite dimensional Teichmüller spaces and generalizations*, Bull. Amer. Math. Soc. **5** (1981), 131–172.

[B8] A. Beurling and L. V. Ahlfors, *The boundary correspondence under quasiconformal mappings*, Acta Math. **96** (1956), 125–142.

[B9] B. Bojarski, *Generalized solutions of a system of first order differential equations of elliptic type with discontinuous coefficients*, Mat. Sb. **43** (1957), 451–503. (Russian)

[B10] M. Brown, *Locally flat embeddings of topological manifolds*, Ann. Math. **75** (1962), 331–341.

[C1] L. Carleson, *The extension problem for quasiconformal mappings*, Contributions to Analysis, Academic Press, New York, 1974, pp. 39–47.

[D1] A. Douady and C. J. Earle, *Conformally natural extension of homeomorphisms of the circle*, Acta Math. (to appear).

[D2] D. Drasin, *The inverse problem of the Nevanlinna theory*, Acta Math. **138** (1977), 83–151.

[E1] C. J. Earle and J. Eells, *A fibre bundle description of Teichmüller theory*, J. Differential Geom. **3** (1969), 19–43.

[E2] C. J. Earle and S. Nag, *Conformally natural reflections in Jordan curves with applications to Teichmüller spaces* (to appear).

[E3] T. Erkama, *Group actions and extension problems for maps of balls*, Ann. Acad. Sci. Fenn. Ser. A I Math. **556** (1973), 1–31.

[F1] C. Fefferman, *Characterizations of bounded mean oscillation*, Bull. Amer. Math. Soc. **77** (1971), 587–588.

[F2] M. H. Freedman and R. Skora, *Strange actions of groups on spheres*, J. Differential Geom. (to appear).

[F3] B. Fuglede, *Extremal length and functional completion*, Acta Math. **98** (1957), 171–219.

[G1] D. B. Gauld and J. Väisälä, *Lipschitz and quasiconformal flattening of spheres and cells*, Ann. Acad. Sci. Fenn. Ser. A I Math. **4** (1978/79), 371–382.

[G2] F. W. Gehring, *Symmetrization of rings in space*, Trans. Amer. Math. Soc. **101** (1961), 499–519.

[G3] _____, *Rings and quasiconformal mappings in space*, Trans. Amer. Math. Soc. **103** (1962), 353–393.

[G4] _____, *Extension of quasiconformal mappings in three space*, J. Analyse Math. **14** (1965), 171–182.

[G5] _____, *Extension theorems for quasiconformal mappings in n-space*, J. Analyse Math. **19** (1967), 149–169.

[G6] _____, *The $L^p$-integrability of the partial derivatives of a quasiconformal mapping*, Acta Math. **130** (1973), 265–277.

[G7] _____, *Quasiconformal mappings*, Complex Analysis and its Applications. II, International Atomic Energy Agency, Vienna, 1976, pp. 213–268.

[G8] _____, *Injectivity of local quasi-isometries*, Comment. Math. Helv. **57** (1982), 202–220.

[G9] _____, *Extension of quasiisometric embeddings of Jordan curves*, Complex Variables Theory Appl. **5** (1986), 245–263.

[G10] F. W. Gehring and G. J. Martin, *Discrete quasiconformal groups* I, Proc. London Math. Soc. (to appear).

[G11] F. W. Gehring and E. Reich, *Area distortion under quasiconformal mappings*, Ann. Acad. Sci. Fenn Ser. A I Math. **388** (1966), 1–15.

[G12] F. W. Gehring and J. Väisälä, *The coefficients of quasiconformality of domains in space*, Acta Math. **114** (1965), 1–70.

[G13] M. Giaquinta, *Multiple integrals in the calculus of variations and nonlinear elliptic systems*, Ann. of Math. Studies, Vol. 105, Princeton Univ. Press, Princeton, N.J., 1983.

[G14] H. Grötzsch, *Über möglichst konforme Abbildungen von schlichten Bereichen*, Ber. Verh. Sächs. Akad. Wiss. Leipzig **84** (1932), 114–120.

[I1] T. Iwaniec, *Extremal inequalities in Sobolev spaces and quasiconformal mappings*, Z. Anal. Anwendungen 1 (1982), 1–16.

[I2] _____, *On $L^p$-integrability in PDE's and quasiregular mappings for large exponents*, Ann. Acad. Sci. Fenn. Ser. A I Math. **7** (1982), 301–322.

[J1] F. John, *Rotation and strain*, Comm. Pure Appl. Math. **14** (1961), 391–413.

[J2] _____, *On quasi-isometric mappings*, II, Comm. Pure Appl. Math. **22** (1969), 265–278.

[J3] F. John and L. Nirenberg, *On functions of bounded mean oscillation*, Comm. Pure Appl. Math. **14** (1961), 415–426.

[J4] P. W. Jones, *Extension theorems for BMO*, Indiana Univ. Math. J. **29** (1980), 41–66.

[K1] M. Kiikka, *Diffeomorphic approximation of quasiconformal and quasisymmetric homeomorphisms*, Ann. Acad. Fenn. Sci. Ser. A I Math. **8** (1983), 251–256.

[K2] I. Kra, *On the Nielsen-Thurston-Bers type of some self-maps of Riemann surfaces*, Acta Math. **146** (1981), 231–270.

[K3] T. Kuusalo, *Quasiconformal mappings without boundary extensions*, Ann. Acad. Sci. Fenn. Ser. A I Math. **10** (1985), 331–338.

[L1] M. A. Lavrentieff, *Sur une classe de représentations continues*, Mat. Sb. **42** (1935), 407–423.

[L2] ____, *Sur un critère différentiel des transformations homéomorphes des domaines à trois dimensions*, Dokl. Akad. Nauk. **20** (1938), 241–242.

[L3] O. Lehto and K. I. Virtanen, *Quasiconformal mappings in the plane*, Springer-Verlag, 1973.

[L4] J. Lelong-Ferrand, *Transformations conformes et quasi-conformes des variétés riemanniennes compactes (Démonstration de la conjoncture de A. Lichnerowicz)*, Acad. Roy. Belg. Cl. Sci. Mém. Collect. **39** (1971), 1–44.

[L5] ____, *Étude d'une classe d'applications liées à des homomorphismes d'algèbres de fonctions, et généralisant les quasi conformes*, Duke Math. J. **40** (1973), 163–186.

[L6] L. G. Lewis, *Quasiconformal mappings and Royden algebras in space*, Trans. Amer. Math. Soc. **158** (1971), 481–492.

[L7] C. Loewner, *On the conformal capacity in space*, J. Math. Mech. **8** (1959), 411–414.

[M1] O. Martio, S. Rickman, and J. Väisälä, *Topological and metric properties of quasiregular mappings*, Ann. Acad. Sci. Fenn. Ser. A I Math. **488** (1971), 1–31.

[M2] B. Mazur, *On embeddings of spheres*, Bull. Amer. Math. Soc. **65** (1959), 59–65.

[M3] D. Menchoff, *Sur une généralisation d'un théorème de M. H. Bohr*, Mat. Sb. **44** (1937), 339–354.

[M4] C. B. Morrey, *On the solutions of quasi-linear elliptic partial differential equations*, Trans. Amer. Math. Soc. **43** (1938), 126–166.

[M5] G. D. Mostow, *Quasi-conformal mappings in n-space and the rigidity of hyperbolic space forms*, Inst. Hautes Études Sci. Publ. Math. **34** (1968), 53–104.

[N1] Z. Nehari, *The Schwarzian derivative and schlicht functions*, Bull. Amer. Math. Soc. **55** (1949), 545–551.

[N2] R. Nirenberg, *On quasi-pseudoconformality in several complex variables*, Trans. Amer. Math. Soc. **127** (1967), 233–240.

[R1] H. M. Reimann, *Functions of bounded mean oscillation and quasiconformal mappings*, Comment. Math. Helv. **49** (1974), 260–276.

[R2] Yu. G. Reshetnyak, *Space mappings with bounded distortion*, Sibirsk. Mat. Zh. **8** (1967), 629–658. (Russian)

[R3] ____, *Liouville's theorem on conformal mappings under minimal regularity assumptions*, Sibirsk. Mat. Zh. **8** (1967), 835–840. (Russian)

[R4] ____, *Stability estimates in Liouville's theorem and the $L^p$-integrability of the derivatives of quasiconformal mappings*, Sibirsk. Mat. Zh. **17** (1976), 868–896. (Russian)

[R5] S. Rickman, *Extension over quasiconformally equivalent curves*, Ann. Acad. Sci. Fenn Ser. A I Math. **436** (1969), 1–12.

[R6] ____, *On the number of omitted values of entire quasiregular mappings*, J. Analyse Math. **37** (1980), 100–117.

[S1] D. Sullivan, *Hyperbolic geometry and homeomorphisms*, Geometric Topology, Academic Press, New York, 1979, pp. 543–555.

[S2] ____, *On the ergodic theory at infinity of an arbitrary discrete group of hyperbolic motions*, Riemann Surfaces and Related Topics: Proceedings of the 1978 Stony Brook Conference, Ann. of Math. Studies, No. 97, Princeton Univ. Press, Princeton, N.J., 1981, pp. 465–496.

[S3] ____, *Quasiconformal homeomorphisms and dynamics I. Solution of the Fatou-Julia problem on wandering domains*, Ann. of Math. **122** (1985), 401–418.

[S4] ____, *Quasiconformal homeomorphisms and dynamics II: Structural stability implies hyperbolicity for Kleinian groups*, Acta Math. **155** (1985), 243–260.

[T1] O. Teichmüller, *Extremale quasikonforme Abbildungen und quadratische Differentiale*, Abh. Preuss. Akad. Wiss. Mat.-Nat. Kl. **22** (1940), 1–197.

[T2] P. Tukia, *On two-dimensional quasiconformal groups*, Ann. Acad. Sci. Fenn. Ser. A I Math. **5** (1980), 73–78.

38

[T3] ____, *A quasiconformal group not isomorphic to a Möbius group*, Ann. Acad. Sci. Fenn. Ser. A I Math. **6** (1981), 149–160.

[T4] ____, *Quasiconformal extension of quasisymmetric mappings compatible with a Möbius group*, Acta Math. **154** (1985), 153–193.

[T5] P. Tukia and J. Väisälä, *Quasiconformal extension from dimension $n$ to $n+1$*, Ann. of Math. **115** (1982), 331–348.

[T6] ____, *Bilipschitz extensions of maps having quasiconformal extensions*, Math. Ann. **269** (1984), 561–572.

[V1] J. Väisälä, *On quasiconformal mappings in space*, Ann. Acad. Sci. Fenn. Ser. A I Math. **298** (1961), 1–36.

[V2] ____, *On quasiconformal mappings of a ball*, Ann. Acad. Sci. Fenn. Ser. A I Math. **304** (1961), 1–7.

[V3] ____, *Lectures on $n$-dimensional quasiconformal mappings*, Lectures Notes in Math., Vol. 229, Springer-Verlag, 1971.

[V4] ____, *A survey of quasiregular maps in $R^n$*, Proc. Internat. Congr. Math. (Helsinki, 1978), Acad. Sci. Fennica, Helsinki, 1980, pp. 685–691.

[V5] ____, *Quasimöbius maps*, J. Analyse Math. **44** (1984/85), 218–234.

MATHEMATICAL SCIENCES RESEARCH INSTITUTE, BERKELEY, CALIFORNIA 94720, USA

UNIVERSITY OF MICHIGAN, ANN ARBOR, MICHIGAN 48109, USA

Quasiconformal Space Mappings
– A collection of surveys 1960–1990
Springer–Verlag (1992), 39–64
Lecture Notes in Mathematics Vol. 1508

# $L^p$–THEORY OF QUASIREGULAR MAPPINGS

Tadeusz Iwaniec

Syracuse University, Syracuse, NY 13210

## Introduction

Historically the origin of quasiconformal mappings is connected with developments of the methods of complex functions. Since the power of this concept was first realized, quasiconformal mappings have engaged the attention of many prominent mathematicians and the theory has been greatly expanded to higher dimensions. The fundamental principle of this theory is to interpolate between diffeomorphisms and homeomorphisms [G4]. One difference between conformal and quasiconformal mappings is that the latter need not be differentiable in the usual sense. However, by a theorem due to A. Mori, F. W. Gehring and J. Väisälä, every quasiconformal mapping $f : \Omega \to \mathbf{R}^n$ is differentiable almost everywhere and its Jacobian determinant $\mathcal{J}(x, f)$ is locally integrable. Moreover, quasiconformality of $f$ can be expressed by the differential inequality

$$(0.1) \qquad |Df(x)|^n \leq K\mathcal{J}(x, f),$$

where $|Df(x)|$ denotes the norm of the differential $Df(x) : \mathbf{R}^n \to \mathbf{R}^n$ and the constant $K \geq 1$ is independent of $x \in \Omega$. Hence quasiconformal mappings can be treated by the methods of measure and integration. This fundamental result allows for further generalizations, due to Yu. G. Reshetnyak, that include non–injective mappings. For the basic results see [BI], [Re5], [Ri2], [Vä] and [V].

To limit the necessary preliminaries we formulate, as a starting point, a slightly more general definition.

For an open subset $\Omega$ of $\mathbf{R}^n$ we shall consider a mapping $f : \Omega \to \mathbf{R}^n$, $f = (f^1, \ldots, f^n)$, of Sobolev class $W^1_{s,\text{loc}}(\Omega, \mathbf{R}^n)$, $1 \leq s < \infty$. Throughout this survey it will be assumed that the Jacobian determinant $\mathcal{J}(x,f) = \det Df(x)$ is non–negative almost everywhere.

**Definition 0.1.** *A mapping* $f \in W^1_{s,\mathrm{loc}}(\Omega, \mathbf{R}^n)$ *is said to be weakly* $K$*–quasiregular,* $1 \leq K < \infty$, *if the dilatation condition*

$$(0.2) \qquad \max_{|\xi|=1} |Df(x)\xi| \leq K \min_{|\xi|=1} |Df(x)\xi|$$

*is satisfied for almost every* $x \in \Omega$. *It is called* $K$*–quasiregular if* $s$ *equals the dimension of the domain, thus* $\mathcal{J}(x,f) \in L^1_{\mathrm{loc}}(\Omega)$. *The smallest number* $K$ *for which* (0.2) *holds will be referred to as the dilatation of* $f$. *If, in addition,* $f$ *is a homeomorphism, then we call it a* $K$*–quasiconformal mapping.*

The objective of this survey is to give some insight into $L^p$–estimates for quasiregular mappings, as well as to state some recent results related to such estimates. In view of the diversity of different approaches it is clearly impossible to expound the subject adequately in a few pages. I wish to devote this article almost exclusively to selected problems concerning quasiregular mappings and certain elliptic equations. One reason, of course, is my personal interest. The other motivation comes from the latest developments [DS], [IM1,3], [I10], that have led to important new results. I found it difficult to give credit where it is due for some of my material, almost none of which is original. I have adopted an informal style, with the hope that readers shall not mistake informality for lack of rigor. I will need to rephrase some of the well-known results.

*Acknowledgement.* This research was supported in part by the U.S. National Science Foundation DMS-9007946.

## 1. Bojarski's theorem

The $L^p$–theory of quasiregular mappings began essentially with the work by B. Bojarski [Bo 1,2]. We recall that $K$–quasiregular mappings on plane domains [A], [LV] are weak solutions of the linear complex Beltrami equation

$$(1.1) \qquad \frac{\partial f}{\partial \bar{z}} = \mu(z)\frac{\partial f}{\partial z},$$

where the complex dilatation $\mu$ is a measurable function such that

$$(1.2) \qquad |\mu(x)| \leq \frac{K-1}{K+1} < 1, \text{ almost everywhere.}$$

We shall assume that $\mu \in L^\infty(\mathbf{C})$ has compact support. One particular solution of (1.1) has the form

$$(1.3) \qquad f(z) = z + \frac{1}{\pi} \iint\limits_{\mathbf{C}} \frac{\omega(\xi)d\sigma(\xi)}{z - \xi},$$

where $\omega = \partial f / \partial \bar{z} \in L^2(\mathbf{C})$ is found from the integral equation

$$(1.4) \qquad \omega - \mu S \omega = \mu.$$

Here $S : L^p(\mathbf{C}) \to L^p(\mathbf{C})$, $1 < p < \infty$, is the familiar complex Hilbert transform defined by the singular integral

$$(1.5) \qquad (S\omega)(z) = -\frac{1}{\pi} \iint\limits_{\mathbf{C}} \frac{\omega(\xi) d\sigma(\xi)}{(z - \xi)^2},$$

also known as the Beurling–Ahlfors transform. The characteristic property is the identity $\partial / \partial z = S \circ \partial / \partial \bar{z}$, connecting the Cauchy–Riemann derivatives.

Equation (1.4), in view of the unitary property of $S$, admits exactly one solution in $L^2(\mathbf{C})$. More importantly, formula (1.3) defines a quasiconformal mapping $f : \mathbf{C} \to \mathbf{C}$, and the other solutions of the Beltrami equation are obtained from this particular one by composing $f$ with analytic functions. This is what we call a Stoilow type factorization of a quasiregular mapping. Very recently we were able to prove such a factorization for solutions of non-uniformly elliptic Beltrami equations [ISv]. $L^p$–estimates for plane quasiregular mappings depend on the norms $A(p) = \|S : L^p(\mathbf{C}) \to L^p(\mathbf{C})\|$, $1 < p < \infty$. Note that

$$(1.6) \qquad a^{-1} \max \{\frac{1}{p-1}, p-1\} \le A(p) \le a \max \{\frac{1}{p-1}, p-1\}$$

for some $a \ge 1$ and that $A(p) = A(q)$ for any Hölder conjugate pair $(q, p)$, $p + q = p \cdot q$.

The goal now is to invert the operator

$$(1.7) \qquad I - \mu S : L^p(\mathbf{C}) \to L^p(\mathbf{C})$$

for $p$ different from 2. Obviously, the inverse exists if $\|\mu\|_\infty A(p) < 1$, that is for $K$ satisfying the condition

$$(1.8) \qquad 1 \le K < \frac{A(p) + 1}{A(p) - 1}.$$

By the Riesz–Thorin convexity theorem we know that the function $A(p)$ is continuous and assumes its minimum value at $p = 2$, $A(2) = 1$.

It is expected that

**Conjecture 1.1.** (See [13]). *For each $p > 1$ we have*

$$A(p) = \max \{\frac{1}{p-1}, p-1\}.$$

In this way B. Bojarski has come to the following conclusion.

**Theorem 1.1.** *Let $(q_K, p_K)$ be the Hölder conjugate pair determined by the equation*

$$(1.9) \qquad A(q_K) = A(p_K) = \frac{K+1}{K-1}, \quad q_K < 2 < p_K.$$

Then every weakly $K$–quasiregular mapping $f \in W^1_{q,\mathrm{loc}}(\Omega, \mathbf{C})$ with $q > q_K$ is quasiregular and belongs to the Sobolev class $W^1_{p,\mathrm{loc}}(\Omega, \mathbf{C})$ for each $p < p_K$.

This theorem allows us to derive various estimates in $L^r$–norms, $r \in (q_K, p_K)$, for a $K$–quasiregular mapping $f \in W^1_{r,\mathrm{loc}}(\Omega, \mathbf{C})$. An important estimate is the following Caccioppoli-type inequality:

$$(1.10) \qquad \iint\limits_{\Omega} |\varphi(z) Df(z)|^r d\sigma(z) \leq C_r(K) \iint\limits_{\Omega} |f(z)|^r |\nabla\varphi(z)|^r d\sigma(z),$$

where $\varphi \in C^\infty_0(\Omega)$ and $q_K < r < p_K$.

This theorem follows easily if we apply the operator $S : L^r(\mathbf{C}) \to L^r(\mathbf{C})$ to the equation

$$(\varphi f)_{\bar{z}} - \mu(z)(\varphi f)_z = f(\varphi_{\bar{z}} - \mu\varphi_z).$$

There are far-reaching consequences of inequality (1.10). For $2 < r < p_K$, in view of the Sobolev imbedding theorem, one finds that $f$ is locally Hölder continuous with exponent $\alpha = 1 - 2/r$. Estimates with $q_k < r < 2$ imply non–trivial removability results for bounded $K$–quasiregular mappings. In Section 8 we discuss this subject in greater generality.

As might be expected, the degree of integrability of $Df$ depends on the regularity of the complex dilatation $\mu(z)$. An interesting application of Fredholm index arises when $\mu$ is assumed to be of vanishing mean oscillation, $\mu \in \mathrm{VMO}(\mathbf{C})$. Recall that the space $\mathrm{VMO}(\mathbf{C})$ is the completion of the space of uniformly continuous functions on $\mathbf{C}$ with respect to the BMO–norm.

We want to outline a proof that when $\mu \in VMO(\mathbf{C})$ the operator $I - \mu S : L^p(\mathbf{C}) \to L^p(\mathbf{C})$ is invertible for all $1 < p < \infty$. For this, we need to consider the family

$$\mathcal{T} = \{T^m : m = 0, \pm 1, \pm 2, \ldots\}$$

of operators $T^m : L^2(\mathbf{C}) \to L^2(\mathbf{C})$ defined by the following Fourier relations

$$(T^m \omega)^{\wedge}(\xi) = (\xi/|\xi|)^m \hat{\omega}(\xi).$$

Of course, $T^o = I$ and $T^{m+k} = T^m \circ T^k$. Thus $\mathcal{T}$ is a group of unitary operators in $L^2(\mathbf{C})$. Notice that $(\xi/|\xi|)^m$ is an $L^p$–multiplier for all $1 < p < \infty$, [S2].

For $m \neq 0$ all $T^m$ are singular integral operators of a convolution type

$$(T^m \omega)(z) = -\frac{m}{2\pi} \iint\limits_{\mathbf{C}} \frac{|z - \xi|^{m-2}\omega(\xi)d\sigma(\xi)}{(z - \xi)^m}.$$

In particular $S = T^2$ and the iterated complex Hilbert transforms $S^m = S \circ S \circ \ldots \circ S = T^{2m-1} \circ T^1$ are easily decomposed into two operators with odd kernels. Applying the familiar method of rotation [CW] to $T^{2m-1}$ and $T^1$ we obtain a bound for the $p$–norms of $S^m$ which grows linearly with respect to $m$, $\|S^m\|_p \leq m \cdot C_p$. Therefore, for each $1 < p < \infty$ and $m$ sufficiently large we can write

$$\|\mu\|^m_\infty \|S^m\|_p < 1.$$

This implies that the operator

$$I - \mu^m S^m : L^p(\mathcal{C}) \to L^p(\mathcal{C})$$

is invertible. On the other hand, our operator $I - \mu S$ is a factor of $I - \mu^m S^m$, modulo a compact operator. Indeed, if we define $P_m = I + \mu S + \mu^2 S^2 + \ldots + \mu^{m-1} S^{m-1}$, then

$$P_m(I - \mu S) = (I - \mu S)P_m = I - \mu^m S^m$$

modulo terms involving the commutator $\mu S - S\mu$ as a factor. We shall now refer to an interesting result of Uchiyama [Uc], which states that the commutator $a\mathcal{R} - \mathcal{R}a$ of a Calderon–Zygmund type operator $\mathcal{R} : L^p(\mathbf{R}^n) \to L^p(\mathbf{R}^n)$ and the multiplication by a function $a \in \mathrm{VMO}(\mathbf{R}^n)$ is compact for all $1 < p < \infty$.

Accordingly, $I - \mu S : L^p(\mathbf{C}) \to L^p(\mathbf{C})$ is a Fredholm operator. What remains is to compute its index. This can be done by performing a continuous deformation of $I - \mu S$ to the identity operator, as follows

$$\mathrm{ind}\,(I - \mu S) = \mathrm{ind}\,(I - t\mu S) = \mathrm{ind}\,I = 0$$

for all $0 \le t \le 1$.

In conclusion, we note that for $p \ge 2$ the kernel of $I - \mu S : L^p(\mathbf{C}) \to L^p(\mathbf{C})$ is trivial, since it is imbedded in $L^2(\mathbf{C})$, on which $I - \mu S$ is a bijection. Therefore $I - \mu S$ is invertible in $L^p(\mathbf{C})$ for all $p \ge 2$. The latter is also true for $1 < p \le 2$, as we see by considering the adjoint operator.

Quasiconformal mappings with complex dilatation of class $\mathrm{VMO}(\Omega)$ seem to be of special interest since they are close to diffeomorphisms. For other recent results concerning Beltrami equation see [Da] and [ISv].

## 2. Estimates in the Sobolev norm $W_n^1(\Omega)$

The theory of quasiregular mappings in dimensions greater than two is non–linear. The very first estimates are derived from an elementary identity for a mapping $f = (f^1, \ldots, f^n)$ of Sobolev class $W_{n,\mathrm{loc}}^1(\Omega; \mathbf{R}^n)$;

(2.1) $$\mathcal{J}(x, f)dx = df^1 \wedge df^2 \wedge df^n = d(f^1 df^2 \wedge \ldots \wedge df^n).$$

This implies rather special properties of the Jacobian determinant. Using Stokes' theorem in a fairly routine manner, we arrive at the inequalities

(2.2) $$\left| \int_\Omega \varphi(x)\mathcal{J}(x, f)dx - \int_\Omega \varphi(x)\mathcal{J}(x, g)dx \right|$$

$$\le \int_\Omega |\nabla\varphi(x)||f(x) - g(x)|(|Df(x)| + |Dg(x)|)^{n-1} dx$$

$$\le \|\nabla\varphi\|_{\infty,\Omega}\|f - g\|_{n,\Omega}(\|Df\|_{n,\Omega} + \|Dg\|_{n,\Omega})^{n-1}$$

for all $f, g \in W_n^1(\Omega, \mathbf{R}^n)$ and $\varphi \in C_o^1(\Omega)$, [BI].

In view of the compactness of the imbedding $W_n^1(\Omega) \subset L^n(\Omega)$ we easily conclude

**Theorem 2.1.** (Yu. G. Reshetnyak [Re1]). *Let* $\{f_j\}$, $j = 1, 2, \ldots$, *be a sequence of mappings* $f_j \in W_n^1(\Omega, \mathbf{R}^n)$ *converging weakly to* $f$. *Then*

$$(2.3) \qquad \lim_{j \to \infty} \int_\Omega \varphi(x) \mathcal{J}(x, f_j) dx = \int_\Omega \varphi(x) \mathcal{J}(x, f) dx$$

*for each* $\varphi \in C_o^\infty(\Omega)$.

If, in addition, all $f_j$'s are $K$–quasiregular then the limit mapping is also $K$–quasiregular. Indeed, for each $\varphi \geq 0$, $\varphi \in C_o^\infty(\Omega)$, we can write

$$\int_\Omega \varphi(x) |Df(x)|^n dx \leq \lim_{j \to \infty} \int_\Omega \varphi(x) |Df_j(x)|^n dx$$

$$\leq K \lim_{j \to \infty} \int_\Omega \varphi(x) \mathcal{J}(x, f_j) dx = K \int_\Omega \varphi(x) \mathcal{J}(x, f) dx.$$

This implies the pointwise estimate (0.1) for the mapping $f$, as desired. Thus we have obtained a weak form of the normal family theorem for $K$–quasiregular mappings.

When $g$ is a constant map, say $g = c \in \mathbf{R}^n$, estimate (2.2) reduces to

$$(2.4) \qquad \int_\Omega \varphi(x) \mathcal{J}(x, f) dx \leq \int_\Omega |\nabla \varphi(x)| |f(x) - c| |Df(x)|^{n-1} dx,$$

for all $f \in W_{n,\text{loc}}^1(\Omega, \mathbf{R}^n)$. Assuming that $f$ is $K$–quasiregular, in view of (0.1), we obtain a Caccioppoli–type inequality

$$(2.5) \qquad \left( \int_\Omega |\varphi Df|^n \right)^{1/n} \leq nK \left( \int_\Omega |f - c|^n |\nabla \varphi|^n \right)^{1/n}.$$

This result, though elementary, has deep consequences. For example, it implies that the sets of zero $n$–capacity are removable under bounded quasiregular mappings [MRV], [Re2]; see Section 8 for more general results.

Let $f \in W_n^1(\Omega, \mathbf{R}^n)$ be an arbitrary mapping with non–negative Jacobian determinant. Inequality (2.4) together with the Poincaré–Sobolev Lemma implies important local estimates. For a cube $Q$ in $\Omega$ and $0 < \sigma < 1$ we denote by $\sigma Q$ the cube with the same center as $Q$ but $\sigma$–times as small. Then, (2.4) applied to the cut–off function of $\sigma Q \subset Q$ yields

$$(2.6) \qquad \frac{1}{|\sigma Q|} \int_{\sigma Q} \mathcal{J}(x, f) dx \leq \frac{C(n)}{\sigma^n (1 - \sigma)} \left[ \frac{1}{|Q|} \int_{|Q|} |Df(x)|^{n^2/(n+1)} dx \right]^{\frac{n+1}{n}}.$$

To simplify the notation we write

$$|h|_{p,E} = \left(\frac{1}{|E|} \int\limits_{E} |h(x)|^p dx\right)^{1/p}$$

and

$$h_E = \frac{1}{|E|} \int\limits_{E} h(x)dx$$

for $1 \leq p < \infty$ and $h \in L^p(E)$. We omit the subscript $p$ if it equals one, so $|h|_E = |h|_{1,E}$.

If $f$ is $K$–quasiregular, in view of (0.1), inequality (2.6) reads as

$$(2.7) \qquad |Df|_{n,\sigma Q} \leq C(n,\sigma)K^{1/n}|Df|_{\frac{n^2}{n+1},Q}$$

for each cube $Q \subset \Omega$, where we notice that $n^2/(n+1) < n$. For $K$–quasiconformal mappings estimates of this kind remain valid with $\sigma = 1$, [G3]. For example,

$$(2.8) \qquad |Df|_{n,Q} \leq C(n,K)|Df|_Q,$$

but in this case the constant $C(n,K)$ may depend on the distance of $Q$ to $\partial\Omega$ as well. O. Martio [M] proved (2.8) for quasiregular mappings with constant $C(n,K)$ depending also on the topological degree of $f$.

We need a definition.

**Definition 2.1.** *A function $h \in L^r_{\text{loc}}(\Omega)$, $0 < r < \infty$, is said to satisfy the weak reverse Hölder inequalities of exponents $0 < q < r$ if there exist $0 < \sigma \leq 1$ and $A \geq 1$ such that*

$$(2.9) \qquad |h|_{r,\sigma Q} \leq A|h|_{q,Q}$$

*for all cubes $Q \subset \Omega$.*

We ignore the term "weak" in case $\sigma = 1$.

## 3. Gehring's lemma

One of the most profound results in multi-dimensional quasiconformal analysis is the higher integrability theorem of F.W. Gehring [G3].

**Theorem 3.1.** *For each $n \geq 2$ and $K \geq 1$ there exists $p = p(n,K) > n$ such that every $K$–quasiconformal mapping $f : \Omega \to \mathbf{R}^n$ belongs to $W^1_{p,\text{loc}}(\Omega, \mathbf{R}^n)$. For a compact subset $F$ of $\Omega$ there is $C = C(n,K,F,\Omega)$ such that*

$$(3.1) \qquad \|Df\|_{p,F} \leq C\|Df\|_{n,\Omega}.$$

The proof rests on the following:

**Lemma 3.1.** (F.W. Gehring). *Suppose that $h \in L^r(\Omega)$, $r > q > 0$, satisfies reverse Hölder inequalities*

$$|h|_{r,Q} \leq A|h|_{q,Q}$$

*for each cube $Q \subset \Omega$, where $A$ is a constant independent of the cube $Q$. Then there exists $p = p(n, r, q, A) > r$ and $B = B(n, r, q, A) \geq 1$ such that $h \in L^p_{\mathrm{loc}}(\Omega)$ and*

$$|h|_{p,Q} \leq B|h|_{q,Q}$$

*for each cube $Q \subset \Omega$.*

Elcrat and Meyers [ME] were the first to notice that Gehring's lemma is also valid for weak reverse Hölder inequalities and as such might be useful for non-linear elliptic equations; see [GM], [BI] and [St] for rigorous proofs and applications to variational problems.

Nowadays the subject has an independent and active life of its own, with many new and even surprising results. Much of the credit goes to several authors [AS], [Boc], [Gi], [FM], [FS], [IN], [I2,5,9], [MS], [Mi], [Sb1,2,3]. The expository versions are given in [Bo3] and [I6].

A very useful refinement of Gehring's Lemma can be phrased as follows:

**Lemma 3.2.** *Let $0 < \sigma \leq 1$, $0 < q < r$, and let $h \in L^r_{\mathrm{loc}}(\Omega)$ satisfy the weak reverse Hölder inequalities*

$$|h|_{r,\sigma Q} \leq A|h|_{q,Q}$$

*for all cubes $Q \subset \Omega$. Then there exists $p = p(n, r, q, A) > r$ and $B = B(n, r, q, \sigma, A)$ such that $h \in L^p_{\mathrm{loc}}(\Omega)$ and*

$$|h|_{p,\sigma Q} \leq B|h|_{q,Q}$$

*for all cubes $Q \subset \Omega$.*

Now we see, in view of (2.7), that Theorem 3.1 remains valid for $K$–quasiregular mappings. There are many other applications of Gehring's lemma, sometimes enriched by important innovations.

The higher integrability theorem allows for the most natural proofs of the basic properties of a quasiregular mapping, such as Hölder continuity, almost everywhere differentiability, Lusin's condition $\mathcal{N}$, strict inequality $\mathcal{J}(x, f) > 0$ a.e. for $f \neq \mathrm{const}$, etc.; see [BI].

## 4. Maximal inequalities

While the applications of singular integrals are limited essentially to linear problems the concept of a maximal function [CF], [FS], [Mu] can be used in the study of non-linear PDE's, including quasiregular mappings; see [BI] and [I4,5]. For the sake of brevity we focus primarily on those aspects which are related to the reverse Hölder inequalities and a few applications in other unfamiliar directions. More detailed treatments are available in [Gu], [S2] and [I9].

In this section $\Omega$ is a fixed cube in $\mathbf{R}^n$, which may degenerate to the entire space $\mathbf{R}^n$. For $h \in L^p(\Omega)$, $1 \le p \le \infty$, the Hardy–Littlewood maximal function related to $\Omega$ is defined by

$$(\mathcal{M}_p h)(x) = \sup \{|h|_{p,Q}; \ x \in Q \subset \Omega\}.$$

The case $p = 1$ is special, so we write $(\mathcal{M}_1 h)(x) = (\mathcal{M}h)(x)$. Hölder's inequality yields $\mathcal{M}_p h \le \mathcal{M}_q h \le \mathcal{M}_\infty h = |h|_{\infty,\Omega}$, whenever $1 \le p \le q \le \infty$. The well-known maximal theorem [W] asserts that $\mathcal{M}_p$ acts as a sublinear operator from $L^r(\Omega)$ into itself for all $r > p$. Moreover,

$$(4.1) \qquad \int_\Omega |\mathcal{M}_p h|^r \le \frac{3^n r 2^r}{r - p} \int_\Omega |h|^r.$$

The constant on the right hand side of (4.1) exhibits the correct asymptotic behavior as $r$ approaches $p$. The proof is based on Vitali's covering lemma. There is a reverse counterpart of this inequality. If $h$ is a measurable function on $\Omega$ we shall make use of its distribution function

$$h_*(t) = |\{x \in \Omega; |h(x)| > t\}|, \ 0 < t < \infty.$$

Recall the elementary Chebyshev inequality

$$h_*(t) \le \frac{1}{t^p} \int_{|h|>t} |h(x)|^p dx.$$

In order to state an inequality in the opposite direction we should replace $h_*$ by the distribution function of $\mathcal{M}_p h$, [BI], [I9].

$$(4.2) \qquad \frac{1}{t^p} \int_{|h|>t} |h(x)|^p \, dx \le 2^n (\mathcal{M}_p h)_*(t)$$

for all $t > |h|_{p,\Omega}$. Here the $L^p$–average of $|h|$ should be understood to be equal to zero if $\Omega = \mathbf{R}^n$. The proof follows from the Calderon–Zygmund decomposition lemma [S2].

Inequality (4.2) implies that if $\mathcal{M}h$ is integrable on $\Omega$ then $h$ belongs to the Zygmund class $L \log L(\Omega)$ [SI]. This is, perhaps, the first indication of why maximal functions can be used to improve the degree of integrability. We need to discuss this case in more detail.

A function $h \in L^1(\Omega)$, $0 < |\Omega| < \infty$, is said to be in the class $L \log L(\Omega)$ if

$$(4.3) \qquad \|h\|_{L \log L(\Omega)} = \frac{1}{|\Omega|} \int_\Omega |h(x)| \log \left(e + \frac{|h(x)|}{|h|_\Omega}\right) dx < \infty,$$

where $e = 2.71\ldots$. It is a good exercise to verify that the above integral expresses an order-preserving norm, that is $|h(x)| \le |g(x)|$ implies $\|h\|_{L \log L(\Omega)} \le \|g\|_{L \log L(\Omega)}$, [IK].

The dual space, denoted by Exp $(\Omega)$, consists of functions $g$ such that

(4.4)
$$\|g\|_{\text{Exp}\,(\Omega)} = \inf_{\lambda>0} \lambda^{-1} \int_{\Omega} e^{\lambda|g(x)|} dx < \infty.$$

Then we have

(4.5)
$$\left| \int_{\Omega} g(x)h(x)dx \right| \leq \|h\|_{L \log L(\Omega)} \|g\|_{\text{Exp}\,(\Omega)}.$$

**Lemma 4.1.** *Suppose* $\mathcal{M}h \in L^1(\Omega)$. *Then* $h \in L \log L(\Omega)$ *and we have*

(4.6)
$$\int_{\Omega} |h| \log \left( e + \frac{|h|}{|h|_\Omega} \right) \leq 2^{n+1} \int_{\Omega} \mathcal{M}h.$$

**Proof.** We shall use (4.2) with $p = 1$. Denote $a = |h|_\Omega$. Integration by parts yields

$$\int_{|h|>a} |h| \log \left( e + \frac{|h|}{a} \right) = - \int_a^\infty \log \left( e + \frac{t}{a} \right) \left( \int_{|h|>t} |h| \right)' dt$$

$$= - \log (1+e) \int_{|h|>a} |h| + \int_a^\infty \frac{1}{t + ae} \left( \int_{|h|>t} |h| \right) dt$$

$$\leq 2^n \int_0^\infty (\mathcal{M}h)_*(t)dt = 2^n \int_{\Omega} \mathcal{M}h.$$

On the other hand, we have a trivial estimate

$$\int_{|h|<a} |h| \log \left( e + \frac{|h|}{a} \right) \leq \log (1+e) \int_{\Omega} |h| \leq 2 \int_{\Omega} \mathcal{M}h.$$

These two inequalities imply (4.6).

Arguments similar to the above lead to another consequence of (4.2).

**Lemma 4.2.** ([BI], [I9]) *For* $r > p \geq 1$ *and* $h \in L^r(\Omega)$ *we have*

(4.7)
$$\fint_{\Omega} |h|^r \leq \left( \fint_{\Omega} |h|^p \right)^{\frac{r}{p}} + \frac{2^n(r-p)}{r} \fint_{\Omega} |\mathcal{M}_p h|^r,$$

*where* $\fint_{\Omega}$ *stands for the integral mean over* $\Omega$. *For* $\Omega = \mathbf{R}^n$ *this reduces to*

(4.8)
$$\int_{\mathbf{R}^n} |h|^r \leq \frac{2^n(r-p)}{r} \int_{\mathbf{R}^n} |\mathcal{M}_p h|^r.$$

Notice that the factor $r - p$, crucial for some applications, appears in both (4.1) and (4.7). For this reason Lemma 4.2 should be regarded as a reverse maximal theorem. To illustrate an application we sketch a proof of a nonhomogeneous version of Gehring's Lemma; see [G].

Suppose that $h \in L^p(\Omega)$ satisfies local estimates

$$（4.9） \qquad |h|_{p,Q} \leq A|h|_{q,Q} + |f|_{p,Q}, \quad p > q > 1,$$

for each cube $Q \subset \Omega$, where $f$ is a given function of class $L^r(\Omega)$, $r > p$. First assume that $h \in L^r(\Omega)$. Thus (4.9) implies a pointwise estimate of the corresponding maximal functions

$$（4.10） \qquad (\mathcal{M}_p h)(x) \leq A(\mathcal{M}_q h)(x) + (\mathcal{M}_p f)(x).$$

It follows from (4.7) and (4.1) that

$$|h|_{r,\Omega}^r \leq |h|_{p,\Omega}^r + \frac{2^{n+r-1}(r-p)}{r}\left[ A^r |\mathcal{M}_q h|_{r,\Omega}^r + |\mathcal{M}_p f|_{r,\Omega}^r \right]$$

$$\leq |h|_{p,\Omega}^r + \frac{4^r 6^n (r-p)}{r-q} A^r |h|_{r,\Omega}^r + \frac{1}{2} \, 4^r 6^n |f|_{r,\Omega}^r.$$

Clearly, for $r = r(n, p, q, A) > p$ sufficiently close to $p$, this yields

$$（4.11） \qquad |h|_{r,\Omega}^r \leq 2|h|_{p,\Omega}^r + 4^r 6^n |f|_{r,\Omega}^r,$$

as desired.

In order to complete the proof of Gehring's Lemma one should get rid of the redundant hypothesis that $h \in L^r(\Omega)$. This can be done via an approximation of $h$ by smooth functions. There is such an approximation that does not violate the reverse Hölder inequalities (4.9); see [BI], [F] and [I9]. On the other hand, the pointwise inequality (4.10) fails to be preserved under any smooth approximation, even if one replaces $A$ by a different constant. That is why the inequalities of maximal functions alone do not guarantee the higher integrability of $h$; an example is due to M. Giaquinta.

## 5. Müller's Theorem

There are many other, sometimes surprising, consequences of the maximal inequalities. Most recently, Stefan Müller [Mü] has discovered a higher integrability property of the Jacobian function. In order to formulate precise estimates and to keep it simple we assume that $\Omega$ is a cube in $\mathbf{R}^n$ and $F = \sigma\Omega$ for some $\sigma \in (1/2, 1)$.

**Theorem 5.1.** (S. Müller) *If the Jacobian determinant* $\mathcal{J}(x) = \mathcal{J}(x, f)$ *of a mapping* $f \in W_n^1(\Omega, \mathbf{R}^n)$ *does not change sign, then it belongs to the Zygmund class* $L \log L(F)$ *and the following uniform bound holds:*

$$（5.1） \qquad \int_F |\mathcal{J}(x)| \log \left( e + \frac{|\mathcal{J}(x)|}{|\mathcal{J}|_F} \right) dx \leq \frac{C(n)}{1-\sigma} \int_\Omega |Df(x)|^n dx.$$

Indeed, (2.5) yields the pointwise estimate

$$(\mathcal{M}\mathcal{J})(x) \leq \frac{C(n)}{1-\sigma}(\mathcal{M}|Df|^{\frac{n^2}{n+1}})(x)$$

for each $x \in F$. Here $\mathcal{M}\mathcal{J}$ stands for the maximal function of $\mathcal{J}$ related to the cube $F$, while on the right hand side the maximal function of $|Df|^{n^2/(n+1)} \in L^{\frac{n+1}{n}}(\Omega)$ is related to the cube $\Omega$. Maximal inequality (4.6) and inequality (4.1) with $p = 1$ and $r = 1+1/n$ lead to the desired estimate.

With the aid of this theorem S. Müller has improved Reshetnyak's result on the weak convergence of Jacobians. We shall formulate a somewhat stronger version of Müller's convergence theorem. To this end, we should observe that the class $C_o^\infty(\Omega)$ is dense in $\exp(\Omega)$— the closure of bounded functions in $\mathrm{Exp}(\Omega)$. A routine approximation argument together with (4.5) and (5.1) gives the following improvement of Theorem 2.1:

**Corollary 5.1.** *Let $f_j \in W_n^1(\Omega, \mathbf{R}^n)$, $j = 1, 2, \ldots$, be mappings with non–negative Jacobian, converging to $f$ weakly in $W_n^1(\Omega, \mathbf{R}^n)$. Then*

$$(5.2) \qquad \lim_{j \to \infty} \int_\Omega \varphi(x)\mathcal{J}(x, f_j)dx = \int_\Omega \varphi(x)\mathcal{J}(x, f)dx$$

*for each test function $\varphi \in \exp(\Omega)$ with compact support.*

There arises now a dual question. What is the minimal degree of integrability of the differential $Df(x)$ of a sense preserving mapping $f : \Omega \to \mathbf{R}^n$ that ensures local summability of the Jacobian determinant $J(x, f) = \det Df(x)$?

In [IS1] we have established the following sharp result.

**Theorem 5.2.** *Let $f : \Omega \to \mathbf{R}^n$ be a mapping whose differential belongs to $L^n \log^{-1}(\Omega)$ and $J(x, f) \geq 0$, almost everywhere.*
*Then $J(x, f) \in L^1_{\mathrm{loc}}(\Omega)$ and*

$$(5.3) \qquad \int_{\sigma Q} J(x, f)dx \leq C(n, \sigma) \int_Q |Df|^n \log^{-1}\left(e + \frac{|Df|}{|Df|_Q}\right)$$

*for every concentric cubes $\sigma Q \subset Q \subset \Omega$, $0 < \sigma < 1$.*

Although we do not pursue the matter here, Theorem 5.1., Corollary 5.1 and Theorem 5.2. relate directly to the existence and regularity problems in non–linear elasticity [AF], [B1,2], [BMu], [DM].

## 6. First order differential system

The development of the analytic theory of quasiregular mappings depends on parallel advances in PDE's. This connection is well understood because of the work of Yu. G. Reshetnyak [Re2,4,5] and many others [BI], [GLM], [I 1,4,5,6,9,10], [IM1], [Ri2]; see also the article of O. Martio in this issue. Most recently a new approach has been developed toward quasiregular mappings and manifolds with measurable conformal structures [DS], [IM1], [I10], producing important new equations. We shall give a very brief sketch of these equations and only hint at their relevance to Riemannian geometry.

Let $G : \Omega \to GL(n)$ be a bounded measurable function with values in the class of symmetric matrices of determinant equal to 1, such that

$$(6.1) \qquad \lambda^{-1}|\xi| \leq \langle G(x)\xi, \xi \rangle^{1/2} \leq \lambda|\xi|$$

for $(x, \xi) \in \Omega \times \mathbf{R}^n$. The scalar product $\langle G(x)\xi, \zeta \rangle$ of tangent vectors $\xi, \zeta \in T_x\Omega \cong \mathbf{R}^n$ gives rise to a measurable metric on $\Omega$. Every $K$–quasiregular mapping $f : \Omega \to \mathbf{R}^n$ induces its own metric tensor $G(x)$ on $\Omega$ (the matrix dilatation of $f$) such that $f$ becomes a solution of the familiar Beltrami equation

$$(6.2) \qquad D^t f(x) Df(x) = \mathcal{J}(x, f)^{2/n} G(x).$$

This is the very first equation of particular relevance to quasiconformal analysis. The dilatation condition (0.2) ensures assumption (6.1) with $\lambda = K^{1-1/n}$. Homeomorphic solutions of (6.2) are simply conformal mappings with respect to the metric $G(x)$ in $\Omega$. The Beltrami system implies several first order differential equations.

Fix an integer $\ell = 1, 2, \ldots, n-1$ and assume that $f \in W^1_{s,\text{loc}}(\Omega, \mathbf{R}^n)$, $f = (f^1, \ldots, f^n)$, $s = \max\{\ell, n - \ell\}$, is weakly $K$–quasiregular. We consider two differential forms on $\Omega$ :

$$(6.3) \qquad \begin{cases} u = & f^\ell df^1 \wedge \ldots \wedge df^{\ell-1}, \\ v = & *f^{\ell+1} df^{\ell+2} \wedge \ldots \wedge df^n, \end{cases}$$

where $*$ denotes the Hodge star operator. Applying the exterior derivative $d : \wedge^{\ell-1}(\Omega) \to \wedge^\ell(\Omega)$ and its formal adjoint operator $d^*$, we find that $du = (-1)^{\ell-1} df^1 \wedge \ldots \wedge df^\ell$ and $d^*v = (-1)^{\ell+1} *df^{\ell+1} \wedge \ldots \wedge df^n$ are differential $\ell$–forms with locally integrable coefficient (regular distributions). Because of an analogy with the harmonic conjugate functions we say that $u$ and $v$ are conjugate to each other.

The metric tensor $G(x) : \mathbf{R}^n \to \mathbf{R}^n$ induces a linear mapping $G_\#(x) : \wedge^\ell(\mathbf{R}^n) \to \wedge^\ell(\mathbf{R}^n)$ of the $\ell$–vectors in $\mathbf{R}^n$; this mapping is represented by a symmetric $\binom{n}{\ell} \times \binom{n}{\ell}$ –matrix whose entries are the $\ell \times \ell$–minors of $G(x)$. Moreover,

$$(6.4) \qquad K^{\frac{\ell(\ell-n)}{n}}|\xi| \leq \langle G_\#(x)\xi, \xi \rangle^{1/2} \leq K^{\frac{\ell(n-\ell)}{n}}|\xi|$$

for all $\ell$–vectors $\xi \in \wedge^\ell(\mathbf{R}^n)$. Associated with $G_\#(x)$ is the non–linear mapping $A : \Omega \times \wedge^\ell(\mathbf{R}^n) \to \wedge^\ell(\mathbf{R}^n)$ defined by

$$(6.5) \qquad A(x, \xi) = \langle G_\#^{-1}(x)\xi, \xi \rangle^{\frac{p-2}{2}} G_\#^{-1}(x)\xi,$$

where $p = \frac{n}{\ell}$. Its inverse with respect to $\xi$ has a similar form

$$(6.6) \qquad A^{-1}(x, \zeta) = \langle G_\#(x)\zeta, \zeta \rangle^{\frac{q-2}{2}} G_\#(x)\zeta,$$

where $q = \frac{n}{n-\ell}$ is the Hölder exponent conjugate to $p$, and $\xi = A(x, \xi)$.

As shown in [IM1] and [I10] the differential forms $u$ and $v$ satisfy the equation

(6.7) $$A(x, du) = d^* v$$

and its equivalent conjugate equation

(6.8) $$A^{-1}(x, d^* v) = du.$$

Notice that if $n = 2\ell$, then $p = q = 2$, so the above equations are linear with respect to $u$ and $v$. These are simply linear relations with bounded measurable coefficients between the $\ell \times \ell$–minors of the differential $Df(x)$, which in the case of $\ell = 1$ and $G(x) = Id$ coincide with the familiar Cauchy–Riemann system.

To illustrate, suppose that

$$Df(x) = \begin{bmatrix} A(x) & B(x) \\ C(x) & D(x) \end{bmatrix}$$

is the Jacobian matrix of a 1-quasiregular mapping $f \in W^1_\ell(\Omega, \mathbf{R}^{2\ell})$, where $A, B, C$ and $D$ are the $\ell \times \ell$–submatrices. Then

(6.9) $$\begin{cases} \det A(x) = \det D(x) \\ \det B(x) = (-1)^\ell \det C(x). \end{cases}$$

The other indentities are obtained from (6.9) by permuting the rows and colums of $Df(x)$, [IM1]. In terms of $u$ and $v$ these relations take a very simple form

(6.10) $$du = d^* v.$$

Applying the Laplace operator we obtain $\Delta du = (dd^* + d^* d)du = dd^* du = dd^* d^* v = 0$. Hence, all the $\ell \times \ell$–minors of $Df(x)$ are harmonic distributions. In this way we have obtained the following sharp generalization of Liouville's theorem [IM1].

**Theorem 6.1.** *Every weakly 1-quasiregular mapping* $f \in W^1_{\ell,\mathrm{loc}}(\Omega, \mathbf{R}^{2\ell})$ *is a Möbius mapping. The exponent $\ell$ of the Sobolev class is the lowest possible for the theorem to be true.*

For other results concerning even–dimensional quasiregular mappings we refer to [IM1], where one finds a strong analogy with Bojarski's approach to quasiregular mappings in the complex plane.

## 7. Variational equations

New thinking in this already highly developed subject comes from non–linear elasticity theory [B 1,2]. Let $G : \Omega \to GL(n)$ denote the matrix dilatation of $f$. We think of a quasiregular mapping $f : \Omega \to \mathbf{R}^n$ as a deformation of an elastic body. In non–linear elasticity the integrals of the form

$$(7.1) \qquad \mathcal{E}[f] = \int_\Omega \langle G^{-1}(x)Df(x), Df(x) \rangle^{\frac{n}{2}} \, dx$$

define the total energy stored in $\Omega$. As a rule we seek absolute minimizers of $\mathcal{E}[f]$ in the Sobolev class $W_n^1(\Omega, \mathbf{R}^n)$ under certain boundary conditions, the so–called equilibrium solutions. It follows from Hadamard's inequality that $\mathcal{E}[f] \geq n \int_\Omega \mathcal{J}(x, f) dx$ for all $f \in W_n^1(\Omega, \mathbf{R}^n)$. Equality occurs if and only if $f$ satisfies Beltrami's equation (6.2). Thus, quasiregular mappings are the absolute minimizers of $\mathcal{E}[f]$ subject to the condition $\int_\Omega \mathcal{J}(x, f) dx = c$, where $c$ is a given number. Although we do not prescribe $f$ on the boundary of $\Omega$, the volume integral actually depends only on the boundary values of $f$. Thus we are justified in writing the Lagrange–Euler system

$$(7.2) \qquad \operatorname{div} A(x, Df) = \sum_{j=1}^n \frac{\partial}{\partial x_j} A^{ij}(x, Df) = 0, \quad i = 1, 2, \ldots, n,$$

where

$$(7.3) \qquad A(x, L) = \langle G^{-1}(x)L, L \rangle^{\frac{n-2}{2}} G^{-1}(x)L$$

for $(x, L) \in \Omega \times GL(\mathbf{R}^n)$.

Let us stress that system (7.2) for a general stationary point of $\mathcal{E}[f]$ is of second order, whereas the absolute minimizers (quasiregular mappings) solve the first order Beltrami system. It is also of particular importance that the components $f^1, f^2, \ldots, f^n$ of an absolute minimizer uncouple system (7.2). Thus, each coordinate function $u = f^i$, $i = 1, 2, \ldots, n$, of a quasiregular mapping satisfies a single equation [Re 2]

$$(7.4) \qquad \operatorname{div} A(x, \nabla u) = 0,$$

called the $A$–harmonic equation. In the case of a 1–quasiregular mapping this reduces to

$$(7.5) \qquad \operatorname{div} (|\nabla u|^{n-2} \nabla u) = 0,$$

which is a special case of a $p$–harmonic equation

$$(7.6) \qquad \operatorname{div} (|\nabla u|^{p-2} \nabla u) = 0, \quad 1 < p < \infty.$$

For a review of the properties of $A$–harmonic functions see the article of O. Martio in this issue.

More second order differential equations relevant to quasiregular mappings can be derived from (6.7) and (6.8) by applying the operators $d^*$ and $d$:

$$(7.7) \qquad d^* A(x, du) = 0 \quad \text{and} \quad dA^{-1}(x, d^* v) = 0.$$

It is not difficult to see that the above equations also arise as variational equations of certain functionals, namely

$$(7.8) \qquad \mathcal{E}_\#^{-1}[u] = \int_\Omega \langle G_\#^{-1}(x)du, du \rangle^{\frac{p}{2}} dx, \quad p = \frac{n}{\ell},$$

and

$$(7.9) \qquad \mathcal{E}_\#[v] = \int_\Omega \langle G_\#(x)d^*v, d^*v \rangle^{\frac{q}{2}} dx, \quad q = \frac{n}{n-\ell},$$

respectively. However, this interpretation is valid only under the assumption that $du \in L^p(\Omega, \wedge^\ell)$ and $d^*v \in L^q(\Omega, \wedge^\ell)$. Weak solutions of (7.7) will be referred to as $A$–harmonic tensors [I10].

A general non–linear Hodge theory has been initiated by L.M. and R.B. Sibner [Si] and K. Uhlenbeck [U].

Applications of $A$–harmonic tensors to quasiregular mappings follow from $L^p$–estimates for the corresponding non–homogenous equations in $\mathbf{R}^n$, such as

$$(7.10) \qquad d^* A(x, du) = d^* h,$$

where $h$ is a given $\ell$-form and $u$ is an unknown $(\ell-1)$-form on $\mathbf{R}^n$. The equation is understood in the sense of distributions. For $h \in L^{\frac{r}{p-1}}(\mathbf{R}^n, \wedge^{\ell+1})$, $\max\{1, p-1\} < r$, a differential $(\ell-1)$-form $u$ is said to be a weak solution of (7.10) if $du \in L^r(\mathbf{R}^n, \wedge^\ell)$ and

$$(7.11) \qquad \int_{\mathbf{R}^n} \langle A(x, du), d\varphi \rangle = \int_{\mathbf{R}^n} \langle h, d\varphi \rangle,$$

where $\varphi$ is an arbitrary $(\ell-1)$-form such that $d\varphi \in L^{\frac{r}{r-p+1}}(\mathbf{R}^n, \wedge^\ell)$. The following estimate seems to be of fundamental importance.

**Theorem 7.1** [I10]. *There exists $\varepsilon = \varepsilon(n, p, \lambda) > 0$ such that*

$$(7.12) \qquad \|du\|_r^{p-1} \le C(n, p, \lambda) \|h\|_{\frac{r}{p-1}}$$

*whenever $p - \varepsilon \le r \le p + \varepsilon$.*

In the case of 1-forms this inequality has already been established in [I5].

Now we can define a non-linear operator

$$T : L^{\frac{p}{p-1}}(\mathbf{R}^n, \wedge^\ell) \cap L^{\frac{r}{p-1}}(\mathbf{R}^n, \wedge^\ell) \to L^p(\mathbf{R}^n, \wedge^\ell) \cap L^r(\mathbf{R}^n, \wedge^\ell)$$

by the rule $T(h) = du$, which in the linear case of 1–forms, corresponding to the Poisson equation

$$\operatorname{div} \nabla u = \operatorname{div} \omega,$$

reduces to the formula

$$T\omega = -R\langle R, \omega \rangle = -R_i(\sum R_k \omega^k), \quad i = 1, 2, \dots, n,$$

for a vector function $\omega = (\omega^1, \ldots, \omega^n) \in L^r(\mathbf{R}^n, \mathbf{R}^n)$. Here $R_1, R_2, \ldots, R_n$ are the Riesz transforms in $\mathbf{R}^n$. Therefore, $T$ should be regarded as a non–linear counterpart of the second order Riesz operator $R\langle R, \cdot \rangle = R \otimes R$; see [IM2].

## 8. Caccioppoli inequality

The natural class in which to consider solutions of the $A$–harmonic equation $d^* A(x, du) = 0$, where $A(x, \xi)$ is defined by (6.5), is the Sobolev space $L^p_{1,\text{loc}}(\Omega, \mathbf{R}^n)$. For such solutions we easily derive a Caccioppoli-type estimate

$$(8.1) \qquad \|\eta du\|_p \le p\lambda^{2p}\|u \wedge d\eta\|_p$$

for each $\eta \in C_o^\infty(\Omega)$. Of course, $u$ can be replaced by any differential form $u - u_o$, where $u_o$ is a closed form. Then, applying the Poincaré–Sobolev lemma yields weak reverse Hölder inequalities for $du$, and by Gehring's lemma higher integrability of $du$ follows.

Estimates analogous to (8.1) but with an exponent $r$ in place of $p$ require much more sophisticated arguments. The point is that for $u \in L_1^r(\mathbf{R}^n, \wedge^{\ell-1})$ and $r < p$ we cannot test (7.11) with $\varphi = \eta u$, because $d\varphi$ fails to be of class $L^{\frac{r}{r-p+1}}(\mathbf{R}^n, \wedge^\ell)$. The $L^r$–estimates for the non–linear operator $T$ (see Theorem 7.1) enable us to derive more general Caccioppoli-type inequalities for quasiregular mappings.

**Theorem 8.1.** ([IM1], [I10]) *There exist exponents $q = q(n, K) < n < p(n, K) = p$ such that if $f \in W^1_{r,\text{loc}}(\Omega, \mathbf{R}^n)$, $r \in [q, p]$, is weakly $K$–quasiregular, then*

$$(8.2) \qquad \|\varphi Df\|_r \le C(n, K)\|\,|\nabla\varphi| f\|_r$$

*for each $\varphi \in C_o^\infty(\Omega)$.*

In even dimensions a quantitative character of the exponents $q(n, K)$ and $p(n, K)$ has been identified, due to precise estimates of the $p$–norms of the signature operator $S$; see [IM2].

Estimates with exponents $r < n$ are of special importance for quasiregular mappings. One consequence of (8.2) is the following regularity theorem [IM1], [I10].

**Theorem 8.2.** *Let $q = q(n, K) < n < p(n, K) = p$ be as in Theorem 8.1. Then every weakly $K$–quasiregular mapping of class $W^1_{q,\text{loc}}(\Omega, \mathbf{R}^n)$ belongs to $W^1_{p,\text{loc}}(\Omega, \mathbf{R}^n)$, hence is $K$–quasiregular.*

According to Theorem 6.1 we have $q(n, 1) = \frac{n}{2}$ for $n$–even. However, the question whether $q(n, 1) = \frac{n}{2}$ in odd dimensions still remains open. Another, perhaps the most important, consequence of the Caccioppoli estimate with $r < n$ is a removability theorem.

A closed set $E \subset \mathbf{R}^n$ is said to be removable under bounded $K$–quasiregular mappings if for every open set $\Omega \subset \mathbf{R}^n$ each bounded $K$–quasiregular mapping $f : \Omega \backslash E \to \mathbf{R}^n$ extends to a $K$–quasiregular mapping on $\Omega$.

**Theorem 8.3** ([IM1], [I10]). *For each dimension $n = 2, 3, \ldots$ and each $K \geq 1$ there is $\varepsilon = \varepsilon(n, K) > 0$ such that every closed set $E \subset \mathbf{R}^n$ of Hausdorff dimension $\dim_H (E) < \varepsilon$ is removable under bounded $K$-quasiregular mappings.*

In even dimensions [IM1] we were able to give reasonable bounds for $\varepsilon(n, K)$ in terms of $n$ and $K$. For other interesting results in this direction see [JV] and [KM]. In the opposite direction, for $n = 3$, nonremovable Cantor sets of arbitrarily small Hausdorff dimension have been constructed very recently by S. Rickman [Ri 1].

**Note.** Caccioppoli estimates are stronger than the weak reverse Hölder inequalities (2.9). The latter follow from (7.2) via Poincaré–Sobolev Lemma, for which we test (7.2) with rather special functions, the cut–off functions corresponding to the concentric cubes $\sigma Q \subset Q$. In the proof of the removability theorem, however, it is necessary to apply (7.2) with $\varphi$ supported in a highly irregular set $\Omega \setminus E$. Caccioppoli estimates with an arbitrary test function $\varphi \in C_o^\infty(\Omega)$ cannot be recovered from the reverse Hölder inequalities. For further applications of Theorem 7.1 see [IS2].

## 9. Sharp inequalities

Let $p(n, K) > n$ denote the least upper bound of all numbers $p \geq n$ such that every $K$-quasiregular mapping $f : \Omega \to \mathbf{R}^n$ belongs to $W_{p,\mathrm{loc}}^1(\Omega, \mathbf{R}^n)$.

**Conjecture 9.1** ([G2]). *For each $n = 2, 3, \ldots$ and $K > 1$ we have*

$$(9.1) \qquad p(n, K) = \frac{nK}{K - 1}.$$

It follows from the Sobolev theorem that every $K$-quasiregular mapping is Hölder continuous with exponent $\alpha = 1 - \frac{n}{p}$, $n < p < p(n, K)$. On the other hand, it is known that the optimal Hölder exponent $\alpha = \alpha(n, K)$ for a $K$-quasiconformal mapping is equal to $1/K$, [G1]. Thus (9.1) reads as the Sobolev relation

$$(9.2) \qquad \frac{1}{K} = \alpha(n, K) = 1 - \frac{n}{p(n, K)}.$$

This observation leads us to make an even stronger conjecture.

**Conjecture 9.2.** *Let $f : \Omega \to \mathbf{R}^n$ be a quasiconformal mapping of Hölder class $C_{\mathrm{loc}}^\alpha(\Omega, \mathbf{R}^n)$, $0 < \alpha \leq 1$. Then, regardless of the dilatation, $f$ belongs to $W_{p,\mathrm{loc}}^1(\Omega, \mathbf{R}^n)$, provided $n < p < \frac{n}{1 - \alpha}$.*

If this is true, it would mean that the Sobolev theorem is reversible for quasiconformal mappings. Here are some convincing arguments for this conjecture to be true.

If a mapping $f : \Omega \to \mathbf{R}^n$ is Hölder continuous with exponent $0 < \alpha \leq 1$, then for each ball $B \subset \Omega$ we obviously have

$$(9.3) \qquad |f(B)| \leq C|B|^\alpha.$$

In general, condition (9.3) does not imply Hölder continuity of $f$, but it does if $f$ is quasiconformal. Moreover, (9.3) implies Morrey's condition

$$\int_B |Df(x)|^n \, dx \le C(n,K) r^{n\alpha}$$

for each ball $B = B(a,r) \subset \Omega$, thus $f \in C^\alpha_{\mathrm{loc}}(\Omega)$. Most likely, employing geometric measure theory (Vitali's lemma, Whitney's decomposition, Calderon–Zygmund lemma) one could replace balls in (9.3) by Borel subsets of $\Omega$. We make a definition.

**Definition 9.1.** *A continuous mapping $f : \Omega \to \mathbf{R}^n$ is said to satisfy the measure distortion inequality with exponent $\alpha \in (0,1]$ if*

$$(9.4) \qquad\qquad |f(E)| \le M|E|^\alpha$$

*for each Borel subset $E \subset \Omega$, where $M$ does not depend on $E$.*

For a quasiconformal mapping $f$, (9.4) implies that

$$\|\mathcal{J}(x,f)\|_{q,\Omega} \le \frac{M}{1-q+q\alpha} |\Omega|^{\alpha - 1 + 1/q}$$

for $1 \le q < \frac{1}{1-\alpha}$. Hence $f \in W^1_p(\Omega)$ for each $n < p < \frac{n}{1-\alpha}$. Following F.W. Gehring and E. Reich [GR] we now rephrase Conjecture 9.1 as follows.

**Conjecture 9.3.** *(Measure Distortion) Let $f : B \to B$ be a $K$–quasiconformal mapping of a ball $B = B(a,r) \subset \mathbf{R}^n$ onto itself, $f(a) = a$. Then*

$$(9.5) \qquad\qquad |f(E)| \le M(n,K)|B|^{1-\frac{1}{k}}|E|^{\frac{1}{k}}$$

*for all measurable subsets $E \subset B$.*

If $E$ is a ball concentric with $B$, we notice that

$$(9.6) \qquad\qquad |f(E)| \le |B|^{1-\frac{1}{k}}|E|^{1/K},$$

with equality occurring for the radial mapping $f(x) = \left(\frac{R}{|x-a|}\right)^{1-\frac{1}{k}}(x-a)+a$; see [IK]. It may very well be that

$$\lim_{K \to 1} M(n,K) = 1;$$

compare with similar results in [FV]. In dimension 2 the problem reduces (equivalently) to an estimate of the Hilbert transform of the characteristic function $\chi_E$. The weak (1.1)–type property of $S$ implies that

$$(9.7) \qquad\qquad \int_{B-E} |S\chi_E| \le a|E| \log \frac{\lambda |B|}{|E|}$$

for some $a \geq 1$ and $\lambda \geq 1$. By using a parametric technique F.W. Gehring and E. Reich [GR] reduced (9.5) to showing that $a = 1$. More precisely, (9.5) holds with $K^a$ in place of $K$ and the number $\lambda$ in (9.7) affects only the constant $M(2, K)$; see also [BaM].

We believe that $a = \lambda = 1$ are the best constants in (9.7), as is true when $E$ is a disk, two disks, an ellipse, annulus or other sets we have known to date [I8], [I.K]. E. Reich has proved that $a \leq 17$ [Rei]. In [IM3] we were able to improve his estimate by showing that $a \leq 7.28\ldots$.

Another way to attack Gehring's conjecture is to compute the $p$-norm of the complex Hilbert transform. As we see from Bojarski's approach, the formula $p(2, K) = \frac{2K}{K-1}$ would follow if

$$(9.8) \qquad \|S\|_p = A(p) = \max\left\{p - 1, \frac{1}{p-1}\right\};$$

see Conjecture 1.1.

In recent years some progress has been made on the best constants in classical inequalities and norms of certain singular integral operators [Ba], [Be], [Bu], [D], [FIP], [I 3,7], [Pe], [Pi]. The trouble with the complex Hilbert transform and other operators with even kernel is due in part to the limitations of certain probabilistic and harmonic analysis methods. Because of the analyticity of the kernel of $S$ the methods of holomorphic functions seem to be most preferable.

Strong motivation for computing the $p$-norms of $S$ and its iterates $S^m$, $m = \pm 1, \pm 2, \ldots$, (see Section 1) comes from another result. In [IM2] we have generalized the familiar method of rotation (valid for operators with odd kernels) for a large class of operators with even kernels. Accordingly their norms can be estimated in terms of the $p$-norms of the operators $S^m$, $m = 1, 2, \ldots$.

At present it seems rather optimistic to compute either of these norms. However, a slightly weaker conjecture having a better chance of being solved is given now. This new conjecture would imply Conjecture 9.1 for $n = 2$ [I3], [L1,2].

**Conjecture 9.4.** *We have*

$$(9.9) \qquad \lim_{p \to 1} (p - 1)A(p) = \lim_{p \to \infty} \frac{1}{p} A(p) = 1.$$

Analogous questions concerning sharp estimates can also be formulated for elliptic-type equations. It is a typical situation in PDE that one wants to assume the least possible regularity of a solution, and then deduce higher regularity. In some respects this problem is dual to our previous question of how far one can improve the regularity.

To illustrate, let $q(n, K) \in (1, n)$ denote the greatest lower bound of the numbers $q \in (1, n]$ such that every weakly $K$-quasiregular mapping $f \in W^1_{q,\mathrm{loc}}(\Omega, \mathbf{R}^n)$ is in fact quasiregular, that is, $f \in W^1_{n,\mathrm{loc}}(\Omega, \mathbf{R}^n)$. Although the precise values of the numbers $q(n, K)$ and $p(n, K)$ are mainly of a theoretical interest they should be regarded as valuable since the open interval $(q(n, K), p(n, K))$ serves as the range of those exponents $s$ for which the $L^s$-theory applies to $K$-quasiregular mappings. We now summarize. Bojarski's result can be restated as

$$(9.10) \qquad q(2, K) \leq q_K < 2 < p_K \leq p(2, K).$$

O. Lehto and K. Virtanen [LV] have shown that

$$(9.11) \qquad \frac{1}{p(2,K)} + \frac{1}{q(2,K)} \geq 1.$$

It is not known whether $p(2,K)$ and $q(2,K)$ are Hölder conjugate.
In higher dimensions we have

$$(9.12) \qquad p(n,K) > n \quad \text{sec [G3], [GM], [I2], [ME]},$$

$$(9.13) \qquad \lim_{K \to \infty} p(n,K) = \infty \quad \text{see [GRe], [Re3], [I 4.5]},$$

$$(9.14) \qquad q(2\ell, K) < 2\ell \quad \text{and} \quad q(2\ell, 1) = \ell, \quad \text{see [IM1]},$$

$$(9.15) \qquad q(n,K) < n \quad \text{see [I10]}.$$

The latter is not easy even in the case of $K = 1$.

Finally I want to mention one more sharp result concerning $p$–harmonic functions in the plane [I Ma].

**Theorem 9.1.** *Every $p$–harmonic function, $1 < p < \infty$, belongs to the Hölder class $C_{\omega;c}^{k;\alpha}(\Omega)$, where the integer $k \geq 1$ and the Hölder exponent $\alpha \in (0,1]$ of the $k^{\text{th}}$–order derivatives are determined by the formula*

$$(9.16) \qquad k + \alpha = \frac{1}{6}\left[ 7 + \frac{1}{p-1} + \sqrt{1 + \frac{14}{p-1} + \frac{1}{(p-1)^2}} \right].$$

*This result is sharp, except for $p = 2$, in which case $u$ is harmonic.*

The following consequence of this formula is surprising. In spite of the fact that the 1–harmonic equation admits rather irregular solutions, the degree of regularity of a $p$–harmonic function increases to infinity as $p$ approaches 1. It is not known whether this phenomenon occurs in every dimension.

## References

[AF]    Acerbi, E. and Fusco, N., Semicontinuity problems in the calculus of variations, Arch. Rational Mech. Anal. 86 (1984), 125–145.

[A]     Ahlfors, L.V., Lectures on quasiconformal mappings, D. van Nostrand Company, Inc., New York, (1966).

[AS]    D'Apuzzo, L. and Sbordone, C., Reverse Hölder inequalities. A sharp result. Rendiconti di Matematica, Ser. VII, 10, (1990), 357–366.

[Ba]    Baernstein, A., Some sharp inequalities for conjugate functions, Proc. Symp. Pure Math. 35 (1979), 409–416.

[BM]    Baernstein, A. and Manfredi, J., Topics in quasiconformal mappings, Topics in modern harmonic analysis, Istituto Nazionale di Alta Mathematica, Roma (1983), 849–862.

[B1]    Ball, J.M., Convexity conditions and existence theorems in nonlinear elasticity, Arch. Rat. Mech. Anal. 63 (1977), 337–403.

[B2]    Ball, J.M., Discontinuous equilibrium solutions and cavitation in nonlinear elasticity, Phil. Trans. Roy. Soc. London (A) 306 (1982), 557–612.

[BMu]   Ball, J.M. and Murat, F., $W^{1,p}$–quasiconvexity and variational problems for multiple integrals, J. Funct. Anal. 58 (1984), 225–253.

[Be]    Beckner, W., Inequalities in Fourier analysis, Annals of Mathematics, 102 (1975), 159–182.

[Boc]   Boccardo, L., An $L^s$–estimate for the gradient of solutions of some strongly nonlinear unilateral problems, Ann. Math. Pura Appl.

[Bo1]   Bojarski, B., Homeomorphic solutions of Beltrami system, Dokl. Akad. Nauk. SSSR 102 (1955), 661–664.

[Bo2]   Bojarski, B., Generalized solutions of a system of differential equations of first order and elliptic type with discontinuous coefficients, Math. Sb. No. 43 (85), (1957), 451–503.

[Bo3]   Bojarski, B., Remarks on stability of reverse Hölder inequalities and quasiconformal mappings, Ann. Acad. Sci. Fenn. Ser. A.I. (1985), 291–296.

[BI]    Bojarski, B. and Iwaniec, T., Analytical foundations of the theory of quasiconformal mappings in $\mathbf{R}^n$, Ann. Acad. Sci. Fenn. Ser. A.I., 8 (1983), 257–324.

[Bu]    Burkholder, D.L., Boundary value problems and sharp inequalities for martingale transforms, The Annals of Probability, vol. 12, No. 3 (1984), 647–702.

[CF]    Coiffman, R.R. and Fefferman, C., Weighted norm inequalities for maximal functions and singular integrals, Studia Math. 51 (1974), 241–250.

[CW]    Coiffman, R. and Weiss, G., Transference methods in analysis, Amer. Math. Soc. CBMS No. 31.

[DM]    Dacorogna, B. and Marcellini, P., Semicontinuitè pour des integrandes polyconvexes sans continuitè des determinants, C.R. Acad. Sci. Paris t 311, ser. 1, (1990), 393–396.

[Da]    David, G., Solutions De L'equation De Beltrami Avec $\|\mu\|_\infty = 1$, Ann. Acad. Sci. Fenn. Ser. A I Math. 13 (1988), 25–70.

[D]     Davis, B., On the weak type (1.1) inequality for conjugate functions and circular symmetrization, Proc. Amer. Math. Soc. 44 (1974), 307–311.

[DS]    Donaldson, S.K. and Sullivan, D.P., Quasiconformal 4-manifolds, Acta Math., 163 (1989), 181–252.

[FS]    Fefferman, C. and Stein, E.M., $H^p$-spaces of several variables, Acta Math. 129 (1972), 137–193.

[FV]    Fehlman, R. and Vuorinen, M., Mori's theorem for $n$-dimensional quasiconformal mappings, Ann. Acad. Sci. Fenn. Ser. A.I. 13 (1988), 111–124.

[FIP]    Figiel, T., Iwaniec, T. and Pelczynski, A., Computing norms and critical exponents of some operators in $L^p$-spaces, Studia Math. 79 (1984), 227–274.

[F]    Fiorenza, A., On some reverse integral inequalities, Atti sem. Mat. Fis. Univ. Modena, XXXVIII (1990), 481–491.

[FM]    Franciosi, M. and Moscariello, G., Higher integrability results, Manuscripta Math. 52 (1985), 151–170.

[FS]    Fusco, N. and Sbordone, C., Higher integrability of the gradient of minimizers of functionals with nonstandard growth conditions, Comm. Pure Applied Math., vol. XLIII (1990), 673–683.

[G1]    Gehring, F.W., Rings and quasiconformal mappings in space, Trans. Amer. Math. Soc. 103 (1962), 353–393.

[G2]    Gehring, F.W., Open problems, Proc. Romanian–Finnish Seminar on Teichmüller Spaces and Quasiconformal Mappings, Romania 1969, page 306.

[G3]    Gehring, F.W., The $L^p$-integrability of the partial derivatives of a quasiconformal mapping, Acta Math. 130 (1973), 265–277.

[G4]    Gehring, F.W., Topics in quasiconformal mappings, Proceedings of the ICM Berkeley (1986), 62–80.

[GR]    Gehring, F.W. and Reich, E., Area distortion under quasiconformal mappings, Ann. Acad. Sci. Fenn. Ser. A.I. 388 (1966), 1–14.

[Gi]    Giaquinta, M., Multiple integrals in the Calculus of Variations and nonlinear elliptic systems, Annals of Math. Stud. No. 105, Princeton Univ. Press 1983.

[GM]    Giaquinta, M. and Modica, G., Regularity results for some classes of higher order nonlinear elliptic systems, J. reine angew. Math., 311/312 (1979), 145–169.

[GLM]    Granlund, S., Lindqvist, P. and Martio, O., Conformally invariant variational integrals, Trans. Amer. Math. Soc., 277 (1983), 43–73.

[GRe]    Gurov, L.G. and Reshetnyak, Yu.G., An analogue of functions with bounded mean oscillation, Sibirsk. Math. J. 17, 3 (1976), 540–546.

[Gu]    de Guzmán, M., Real Variable Methods in Fourier Analysis, Mathematics Studies 46, North–Holland, Amsterdam, New York and Oxford (1981).

[I1]    Iwaniec, T., Regularity theorems for the solutions of p.d.e. related to quasiregular mappings in several variables, Preprint of Polish Acad. Sciences (Habilitation Thesis), (1978), 1–45.

[I2]    Iwaniec, T., Gehring's reverse maximal function inequality, Proc. International Conference on Approximation and Function Spaces, August 1979, Gdańsk. Edited by Z. Ciesielski, 294–305.

[I3]    Iwaniec, T., Extremal inequalities in Sobolev spaces and quasiconformal mappings, Zeitschrift für Analysis und ihre Anwendungen Bd. 1 (6), (1982), 1–16.

[I4]  Iwaniec, T., On $L^p$–integrability in p.d.e. and quasiregular mappings for large exponents, Ann. Acad. Sci. Fenn. Ser. A.I. 7, (1982), 301–322.

[I5]  Iwaniec, T., Projections onto gradient fields and $L^p$–estimates for degenerated elliptic operators, Studia Math., 75, (1983), 293–312.

[I6]  Iwaniec, T., Some aspects of p.d.e. and quasiregular mappings, Proceedings of the ICM Warsaw (1983), 1193–1208.

[I7]  Iwaniec, T., The best constant in a BMO–inequality for the Beurling–Ahlfors transform, Michigan Math. J. 33 (1986), 387–394.

[I8]  Iwaniec, T., Hilbert transform in the complex plane and area inequalities for certain quadratic differentials, Michigan Math. J. 34 (1987), 407–434.

[I9]  Iwaniec, T., Lectures on quasiconformal mappings at Syracuse University, Notes (1987–88), 1–426.

[I10] Iwaniec, T., p–harmonic tensors and quasiregular mappings, Ann. Math. (to appear).

[IK]  Iwaniec, T. and Kosecki, R., Sharp estimates for complex potentials and quasiconformal mappings, preprint of Syracuse University, 1–69.

[IM1] Iwaniec, T. and Martin, G., Quasiregular mappings in even dimensions, Mittag–Leffler Institute, Report No. 19, 1989/90, Acta Math. (to appear).

[IM2] Iwaniec, T. and Martin, G., The Beurling–Ahlfors transform in $R^n$ and related singular inegrals, Preprint of Inst. Hautes Etudes Sci., 1990.

[IM3] Iwaniec, T. and Martin, G., Quasiconformal mappings and capacity, Indiana Univ. Math. J., 40 No. 1 (1991), 101–122.

[IMa] Iwaniec, T. and Manfredi. J., Regularity of p–Harmonic functions on the plane, Revista Mathemática Iberoamericana, Vol. 5, No. 1 (1989), 1–19.

[IN]  Iwaniec, T. and Nolder, C., The Hardy–Littlewood inequality for quasiregular mapping in certain domains in $R^n$, Ann. Acad. Sci. Fenn. Ser. A.I. vol. 10 (1985), 267–282.

[IS1] Iwaniec, T. and Sbordone, C., On the integrability of the Jacobian under minimal hypotheses, Arch. Rat. Mech. Anal. (to appear).

[IS2] Iwaniec, T. and Sbordone, C., Weak minima of variational integrals, in preparation.

[ISv] Iwaniec, T. and Sverák, V., On mappings with integrable dilatation, (to appear).

[JV]  Järvi, P. and Vuorinen, M., Self–similar Cantor sets and quasiregular mappings, J. reine angew. Math. 424 (1992), 31-45.

[KM]  Koskela, P. and Martio, O., Removability theorems for quasiregular mappings, Ann. Acad. Sci. Fenn. Ser. A.I. vol. 15 (1990), 381–399.

[L1]  Lehto, O., Remarks on the integrability of the derivatives of quasiconformal mappings, Ann. Acad. Sci. Fenn. Ser. A.I. 371 (1965), 1–8.

[L2]  Lehto, O., Quasiconformal mappings and singular integrals, Symposia Mathematica, vol. XVIII, Academic Press London (1976), 429–453.

[LV]  Lehto, O. and Virtanen, K., Quasiconformal mappings in the plane, Second Edition, Springer–Verlag, New York – Heidelberg 1973.

[MS]  Marcellini, P. and Sbordone, C., On the existence of minima of multiple integrals of the Calculus of Variations, J. Math. Pures Appl. 62 (1983), 1–9.

[M]   Martio, O., On the integrability of the derivatives of a quasiregular mapping, Math. Scand. 35 (1974), 43–48.

[MRV] Martio, O., Rickman, S. and Väisälä, J., Distortion and singularities of quasiregular mappings, Ann. Acad. Sci. Fenn. Ser. A.I. 465 (1970), 1–13.

[ME] Meyers, N. and Elcrat, A., Some results on regularity for solutions of nonlinear elliptic systems and quasiregular functions, Duke Math. J., vol. 42 (1) (1975), 121–136.

[Mi] Migliaccio, L., Reverse Hölder from reverse Jensen inequalities, An International Workshop, Capri, Sept. 17–20, 1990, 129–134.

[Mu] Muckenhoupt, B., Weighted norm inequalities for the Hardy maximal function, Trans. Amer. Math. Soc. 165 (1972), 207–226.

[Mü] Müller, S., Higher integrability of determinants and weak convergence in $L^1$, J. reine angew. Math. 412 (1990), 20-34.

[Pe] Pelczynski, A., Norms of classical operators in function spaces, Colloque Laurent Schwartz, Astérisque 131 (1985), 137–162.

[Pi] Pichorides, S.K., On the best values of the constants in the theorems of M. Riesz, Zygmund and Kolmogorov, Studia Math. 44 (1972), 165–179.

[Rei] Reich, E., Some estimates for the two–dimensional Hilbert transform, J. Analyse Math., vol. XVIII (1967), 279–293.

[Re1] Reshetnyak, Yu.G., On the stability of conformal mappings in multidimensional spaces, Siberian Math. J. 8 (1967), 65–85.

[Re2] Reshetnyak, Yu.G., On extremal properties of mappings with bounded distortion, Sib. Math. J. 10 (1969), 1300–1310.

[Re3] Reshetnyak, Yu.G., Stability estimates in Liouville's theorem and the $L^p$–integrability of the derivatives of quasiconformal mappings, Sib. Math. J. 17 (1976), 868–896.

[Re4] Reshetnyak, Yu.G., Differentiability properties of quasiconformal mappings and conformal mappings of Riemannian spaces, Sibirsk. Math. J. 19 (1978), 1166–1183.

[Re5] Reshetnyak, Yu.G., Space mappings with bounded distortion, Trans. Math. Monographs, Amer. Math. Soc., Vol. 73, 1989.

[Ri1] Rickman, S., Nonremovable Cantor sets for bounded quasiregular mappings, Mittag–Leffler Institute, Report No. 42, 1989/90.

[Ri2] Rickman, S., Quasiregular Mappings, Springer–Verlag, (to appear).

[Sb1] Sbordone, C., Rearrangement of functions and reverse Hölder inequalities, Ennio De Giorgi Colloquium Res. Notes in Math., Pitman, 125 (1985), 139–148.

[Sb2] Sbordone, C., On some integral inequalities and their applications to the Calculus of Variations, Boll. Unione Mat. Ital. (6), 5 (1986), 73–94.

[Sb3] Sbordone, C., Rearrangement of functions and reverse Jensen inequalities, Proc. of Symposia in Pure Math. vol. 45 (1986), Part 2, 325–329.

[Si] Sibner, L.M. and Sibner, R.B., A non–linear Hodge de Rham theorem, Acta Math., 125, (1970), 57–73.

[S1] Stein, E.M., Note on the class $L \log L$, Studia Math. 32 (1969), 305–310.

[S2] Stein, E.M., Singular integrals and differentiability properties of functions, Princeton Univ. Press, Princeton, N.J., 1970.

[St] Stredulinsky, E.W., Higher integrability from reverse Hölder inequalities, Indiana Univ. Math. J. 29, 3 (1980), 408–417.

[Uc]  Uchiyama, A., On the compactness of operators of Hankel types, Tôhoku Math. J. 30, (1978), 163–171.

[U]   Uhlenbeck, K., Regularity for a class of non–linear elliptic systems, Acta Math., 138 (1977), 219–240.

[Vä]  Väisälä, J., Lectures on n-dimensional quasiconformal mappings, Lecture Notes in Mathematics, 229, Springer–Verlag 1971.

[V]   Vuorinen, M., Conformal geometry and quasiregular mappings, Lecture Notes in Mathematics, 1319, Springer–Verlag 1988.

[W]   Wiener, N., The ergodic theorem, Duke Math. J., 5, (1939), 1–18.

[Wik] Wik, I., A comparison for the integrability of $f$ and $Mf$ with that of $f^{\#}$, Preprint Series No. 2 (1983), Dept. Math., University of Umeå, Sweden.

Quasiconformal Space Mappings
– A collection of surveys 1960–1990
Springer–Verlag (1992), 65–79
Lecture Notes in Mathematics Vol. 1508

# PARTIAL DIFFERENTIAL EQUATIONS
# AND QUASIREGULAR MAPPINGS

Olli Martio

Department of Mathematics, University of Jyväskylä, Finland

## 1. Introduction

Plane quasiregular mappings form a generalization of analytic functions of one complex variable. Grötzsch introduced regular quasiconformal mappings in 1928. In modern terms, a regular quasiconformal mapping is a continuously differentiable homeomorphic quasiregular mapping; its inverse is usually assumed to be continuously differentiable as well. The early research was directed to problems of a function-theoretic character. The analytic definition of quasiconformality appeared in a paper of Morrey [Mo] in 1938. He studied homeomorphic solutions of a Beltrami system, which is a perturbed form of the Cauchy–Riemann equations. These solutions are understood in the weak sense, and the solutions agree with quasiconformal mappings in modern terminology. This connection remained unnoticed for almost twenty years. Morrey's goal was to determine the class of harmonic morphisms for a non-smooth second order elliptic partial differential equation in the plane. We consider this problem in Sections 5 and 6. In general, the plane situation is understood rather well; see [BJS], [LV], [Ve].

The systematic study of quasiconformal mappings in space started in the 1960's, although these mappings had already been considered by Lavrentiev in 1938. Again the research was first directed to geometric problems of a function-theoretic nature. Studies in quasiconformal mappings have sometimes preceded the corresponding studies in partial differential equations. In particular, the solution, due to Gehring [G], of the $L^p$-integrability problem, $p > n$, of the derivative of a quasiconformal mapping has provided a useful tool in the theory of partial differential equations and systems. However, the applications have been mostly in the other direction.

The study of non-homeomorphic quasiconformal mappings, called quasiregular mappings, in space began in the late 1960's. The classical relation between harmonic and analytic functions was extended by Reshetnyak [R] to quasiregular mappings and solutions of certain partial differential equations. This also provided an extension of Morrey's work from plane to space. In space these equations are always non-linear.

In constrast with the two-dimensional case the theory of partial differential equations was employed at an early stage to obtain the basic topological and metric properties of space quasiregular mappings.

The construction of a quasiconformal or quasiregular mapping is often a geometric task. Since component functions of a quasiregular mapping are solutions of an elliptic second order partial differential equation, this also provides a geometric method obtaining such solutions. For an application of this method see [M2].

Partial differential equations important in the theory of quasiregular mappings are introduced in Section 2 and, after a survey of quasiregular mappings in Section 4, the connection is exploited in Section 5. Since the most interesting partial differential equations are Euler equations of a variational integral, we consider variational integrals in Section 3 and an approach based on variational extremals is included in Section 5. The behavior of superharmonic functions under quasiregular mappings is also studied in Section 5. General harmonic morphisms are considered in Section 6, and Section 7 is devoted to the applications of $A$-harmonic measures to the theory of quasiregular mappings.

## 2. Equations $\nabla \cdot A = 0$ and their solutions

We consider mappings $A \colon \mathbf{R}^n \times \mathbf{R}^n \to \mathbf{R}^n$ satisfying the following assumptions for some $0 < \alpha \le \beta$:

(2.1) $\quad\quad\begin{aligned} x \mapsto A(x,h) &\quad \text{is measurable for all } h \in \mathbf{R}^n \text{ and} \\ h \mapsto A(x,h) &\quad \text{is continuous for a.e. } x \in \mathbf{R}^n, \end{aligned}$

for all $h \in \mathbf{R}^n$ and a.e. $x \in \mathbf{R}^n$

(2.2) $$A(x,h) \cdot h \ge \alpha\,|h|^n,$$

(2.3) $$|A(x,h)| \le \beta\,|h|^{n-1},$$

(2.4) $$\big(A(x,h_1) - A(x,h_2)\big) \cdot (h_1 - h_2) > 0, \quad h_1 \ne h_2,$$

and

(2.5) $$A(x,\lambda h) \equiv |\lambda|^{n-2}\,\lambda A(x,h); \quad \lambda \in \mathbf{R}.$$

The mapping $A$ generates the second order partial differential equation $\nabla \cdot A\big(x, \nabla u(x)\big) = 0$, which is of divergence type. To be more precise, let $G$ be a domain (or an open set) in $\mathbf{R}^n$. Then a function $u \in C(G) \cap \operatorname{loc} W_n^1(G)$ is a solution of the equation

(2.6) $$\nabla \cdot A\big(x, \nabla u(x)\big) = 0$$

induced by the operator $A$ if

(2.7) $$\int_G A\big(x, \nabla u(x)\big) \cdot \nabla \varphi(x)\,dm(x) = 0$$

for all $\varphi \in C_0^\infty(G)$. Here $\operatorname{loc} W_n^1(G)$ is the Sobolev space whose functions are locally $n$-integrable and have first locally $n$-integrable partial distributional derivatives; $\nabla u(x) = (\partial_1 u(x), \dots, \partial_n u(x))$ is the gradient of $u$. Solutions $u$ of (2.6) are called $A$-harmonic.

Note that we require a solution $u$ of (2.6) to be continuous. This is actually a superfluous assumption since if $u \in \operatorname{loc} W_n^1(G)$ satisfies (2.7), then $u$ can be redefined on a set of measure zero in such a way that the new function is continuous.

**2.8. Example.** The mapping $A(x, h) = |h|^{n-2}h$ satisfies assumptions (2.1)–(2.5). The corresponding equation is the $n$-harmonic equation

$$(2.9) \qquad \nabla \cdot (|\nabla u|^{n-2} \nabla u) = 0.$$

In the plane ($n = 2$) this equation reduces to the ordinary Laplace equation $\Delta u = 0$. It is interesting to note that the solutions of (2.9) for $n \geq 3$ need not be $C^2$, although they belong to the class $C^{1,\alpha}$.

**2.10. Properties of $A$-harmonic functions.** In general, an $A$-harmonic function need not be much more regular than an arbitrary function in $C(G) \cap \operatorname{loc} W_n^1(G)$. However, $A$-harmonic functions are locally Hölder continuous in $G$ and they satisfy Harnack's inequality: If $u \geq 0$ is $A$-harmonic in $G$, then

$$\sup_C u \leq c \inf_C u$$

where $C$ is compact in $G$ and $c = c(n, p, \alpha, \beta, C) < \infty$. Another important property is Harnack's principle: If $(u_i)$ is an increasing sequence of $A$-harmonic functions in a domain $G$, then the function $u = \lim u_i$ is either $A$-harmonic or identically $+\infty$ in $G$. For these results see [**S**], [**GLM1**] and [**HKM2**].

The most important property of the equations $\nabla \cdot A = 0$ is the unique solvability of the Dirichlet problem in regular domains, say balls: If $B$ is a ball and if $f \in C(\partial B)$, then there is a unique $A$-harmonic function $u \in C(\bar{B})$ in $B$ with $u = f$ in $\partial B$. For $f \in C^\infty(\mathbb{R}^n)$ this result can be proved by means of the theory of monotone operators and the general case follows by approximation, Harnack's principle, and suitable boundary estimates. In fact, necessary and sufficient boundary conditions for the solvability of the Dirichlet problem for equation (2.6) are known; see [**Maz**] and [**LM**].

The strict monotonicity condition (2.4) yields the following result: If $G$ is a bounded open set and if $u_1, u_2 \in C(\bar{G})$ are $A$-harmonic in $G$, then $u_1 \leq u_2$ in $\partial G$ implies $u_1 \leq u_2$ in $G$. To prove this, assume that $D = \{x \in G : u_1(x) > u_2(x)\}$ is non-empty. Then for some $\varepsilon > 0$, $D_\varepsilon = \{x \in G : u_1(x) > u_2(x) + \varepsilon\}$ is also non-empty. Now $D_\varepsilon$ is open and $D_\varepsilon \Subset G$. Furthermore, $u_1 = u_2 + \varepsilon$ in $\partial D_\varepsilon$ and the function $\varphi = u_1 - u_2 - \varepsilon$ belongs to $W_{n,0}^1(D_\varepsilon)$; the functions in this subspace of $W_n^1(D_\varepsilon)$ can be approximated by $C_0^\infty(D_\varepsilon)$-functions in the Sobolev norm of $W_n^1(D_\varepsilon)$. This means that $\varphi$ can be used in (2.7) to test the $A$-harmonicity of $u_1$ and $u_2$ in $D_\varepsilon$, i.e.

$$\int_{D_\varepsilon} A(x, \nabla u_i) \cdot \nabla \varphi \, dm = 0, \quad i = 1, 2.$$

Since $\nabla \varphi = \nabla u_1 - \nabla u_2$, we obtain

$$\int_{D_\varepsilon} (A(x, \nabla u_1) - A(x, \nabla u_2)) \cdot (\nabla u_1 - \nabla u_2) \, dm = 0.$$

By (2.4), $\nabla u_1 = \nabla u_2$ a.e. in $D_\varepsilon$, and this means that $u_1 - u_2 = \mathrm{const.}$ in each component of $D_\varepsilon$. But $u_1 - u_2 = \varepsilon$ in $D_\varepsilon$ because $\varphi \in W_{n,0}^1(D_\varepsilon)$; hence $D_\varepsilon = \emptyset$, a contradiction. Thus $A$-harmonic functions are order-preserving.

Both the unique solvability and the order preservation are needed in the construction of the theory of $A$-superharmonic functions. The theory applies to more general equations than (2.6). This theory is non-linear whenever equation (2.6) is non-linear, however, in many respects it is similar to the classical potential theory; see [**HKM2**].

**2.11. A-superharmonic functions.** Let $G$ be a domain in $\mathbf{R}^n$ and let $A$ be a mapping satisfying (2.1)–(2.5). A lower semicontinuous (lsc) function $v: G \to R \cup \{\infty\}$ is *A-superharmonic* if $v \not\equiv \infty$ and if $v$ satisfies the $A$-comparison principle, i.e. if for each domain $D \Subset G$ and each function $u \in C(\bar{D})$, $A$-harmonic in $D$, $u \le v$ in $\partial D$ implies $u \le v$ in $D$. A function $v: G \to R \cup \{-\infty\}$ is *A-subharmonic* if $-v$ is $A$-superharmonic.

It is well known (see e.g. [**Ra**]) that for $A(x, h) = h$ the above definition can be used to characterize ordinary superharmonic functions. Another classical definition is the mean value inequality: A lsc function $v: G \to R \cup \{\infty\}$ is superharmonic if and only if

$$(2.12) \qquad v(x_0) \ge \frac{1}{m\big(B(x_0, r)\big)} \int_{B(x_0, r)} v \, dm$$

for each ball $B(x_0, r) \subset G$. Here the volume integral can also be replaced by an $(n-1)$-dimensional integral over $\partial B(x_0, r)$. It is clear that the characterization (2.12) has no counterpart for general operators $A$, the main obstacle being that $v_1 + v_2$ need not be $A$-superharmonic for two $A$-superharmonic functions $v_1$ and $v_2$. Note that condition (2.12) is linear. Observe that the definition for $A$-superharmonicity does not immediately imply that $A$-superharmonicity is a local property.

We mention some properties of $A$-superharmonic functions:

(a) If $v_1$ and $v_2$ are $A$-superharmonic, then $\min(v_1, v_2)$ and $\lambda v_1 + \mu$, $\lambda \ge 0$, $\mu \in \mathbf{R}$, are $A$-superharmonic.

(b) If $(v_i)$ is an increasing sequence of $A$-superharmonic functions, then $v = \lim v_i$ is either $A$-superharmonic or $v \equiv \infty$.

Perron's celebrated method also works for equations (2.6). This makes it possible to formulate and solve the most general boundary value problem and to form special solutions like the $A$-harmonic measure, see Section 7. Also the balayage method is available in the potential theory formed by $A$-harmonic and $A$-superharmonic functions; see [**HKM2**].


**2.13. A-supersolutions.** There is an intermediate class between $A$-harmonic and $A$-superharmonic functions. A function $v \in \operatorname{loc} W_n^1(G)$ is an *A-supersolution* if

$$(2.14) \qquad \int_G A\big(x, \nabla v(x)\big) \cdot \nabla \varphi(x) \, dm(x) \ge 0$$

for all $\varphi \in C_0^\infty(G)$, $\varphi \ge 0$.


**2.15. Remarks.** (a) An $A$-supersolution $v$ differs from a solution (an $A$-harmonic function) in three aspects: (i) $v$ need not be continuous ($v$ is actually defined as an equivalence class in $\operatorname{loc} L^n(G)$ and may fail to be even lsc), (ii) (2.14) holds as an inequality, and (iii) (2.14) holds only for $\varphi \ge 0$. Note that $A$-harmonic functions could be defined initially without continuity but these functions can be proved to be continuous (and even Hölder continuous) after a redefinition on a set of measure zero.

(b) Let $A(x, h) = h$, i.e. we consider the ordinary plane harmonic case. If $v$ is superharmonic, then $-\Delta v$ can be interpreted as a measure, in particular

$$\int_G v \Delta \varphi \, dm \le 0$$

for all $\varphi \in C_0^\infty(G)$, $\varphi \geq 0$. The converse of this fact is the Riesz representation theorem for superharmonic functions. These results have no direct counterparts for $A$-superharmonic functions; there are no Green's formulas for operators $A$. However, there is a constructive method to overcome these difficulties.

The next properties show the relations between $A$-superharmonic functions and $A$-supersolutions. For the proofs see [GLM1], [HK1] and [HKM2].

(a) Let $v$ be an $A$-supersolution. Then $v_1 = \mathrm{ess\,lim}\, v$ is $A$-superharmonic and $v_1 = v$ a.e. Conversely, each $A$-superharmonic function $v \in \mathrm{loc}\, W_n^1(G)$ is an $A$-supersolution.

(b) Suppose that $v \in C(G)$. Then $v$ is an $A$-supersolution if and only if $v$ is $A$-superharmonic. More generally, for lsc functions that are bounded above the classes of $A$-supersolutions and $A$-superharmonic functions coincide.

(c) A function $v\colon G \to R \cup \{\infty\}$ is $A$-superharmonic if and only if for each domain $D \Subset G$ there is an increasing sequence $v_i \in C(\bar{D}) \cap W_n^1(D)$ of $A$-superharmonic functions (or $A$-supersolutions) such that $v = \lim v_i$.

The above results show that $A$-superharmonic functions can be obtained as limits of increasing sequences of regular $A$-supersolutions. A method of proving properties (a)–(c) is to consider obstacle problems. The obstacle method gives a constructive approach to $A$-supersolutions. The ideas here come from Calculus of Variations and this method is more straightforward if equation (2.6) is the Euler equation of a variational integral; see the next section.

## 3. Variational integrals

In classical potential theory harmonic functions minimize the Dirichlet integral. Equations (2.6) need not be Euler equations of any reasonable variational integral. However, many important equations arise this way. Suppose that a function $F\colon \mathbf{R}^n \times \mathbf{R}^n \to \mathbf{R}$ satisfies (2.1) and that for some $0 < \alpha' \leq \beta' < \infty$

$$(3.1) \qquad \alpha'|h|^n \leq F(x,h) \leq \beta'|h|^n,$$
$$(3.2) \qquad h \mapsto F(x,h) \text{ is strictly convex and differentiable,}$$

and

$$(3.3) \qquad F(x, \lambda h) = |\lambda|^n F(x,h), \quad \lambda \in \mathbf{R},$$

for all $h \in \mathbf{R}^n$ a.e. Then $F$ induces a variational integral

$$(3.4) \qquad I_F(u) = \int_G F(x, \nabla u(x))\, dm(x);$$

here $u \in W_n^1(G)$. A function $u \in C(G) \cap \mathcal{F}_{u_0}$ is an *extremal of $I_F$ with boundary values* $u_0 \in W_n^1(G)$ if

$$I_F(u) = \inf_{v \in \mathcal{F}_{u_0}} I_F(v)$$

where

$$\mathcal{F}_{u_0} = \{v \in W_n^1(G)\colon v - u_0 \in W_{n,0}^1(G)\}.$$

Here $W_{n,0}^1(G)$ is the closure of $C_0^\infty(G)$ in $W_n^1(G)$. An extremal $u$ is a solution of the Euler equation induced by $F$:

$$(3.5) \qquad\qquad \nabla \cdot \nabla_h F(x, \nabla u(x)) = 0;$$

here $\nabla_h F$ is the gradient of $h \mapsto F(x, h)$ (cf. (3.2)). It is not difficult to see that $A(x, h) = \nabla_h F(x, h)$ satisfies (2.1)–(2.5); see [**GLM1**]. For example, $F(x, h) = |h|^n$ induces the $n$-harmonic equation and, in particular, for $n = 2$ the Dirichlet integral has the Laplace equation as an Euler equation. Other important examples of variational kernels $F$ satisfying (3.1)–(3.3) are of the form

$$F(x, h) = |\theta(x) h|^n$$

where, for each $x \in \mathbf{R}^n$, $\theta(x)$ is a symmetric linear map of $\mathbf{R}^n$ with $\theta(x) h \cdot h \geq \alpha |h|^2$ and $|\theta(x) h| \leq \beta |h|$ for a.e. $x \in \mathbf{R}^n$.

We say that $u$ is a *free extremal* of (3.4) if $u \in C(G) \cap \operatorname{loc} W_n^1(G)$ and if for each domain $D \Subset G$, $u$ is an extremal of $I_F$ in $D$ with boundary values $u$. Now extremals and free extremals are related to $A$-harmonic functions, $A = \nabla_h F(x, h)$, in the following way:

(a) $u \in C(G) \cap W_n^1(G)$ is $A$-harmonic in $G$ with $u - u_0 \in W_{n,0}^1(G)$ if and only if $u$ is an extremal of $I_F$ with boundary values $u_0$.

(b) $u$ is a free extremal in $G$ if and only if $u$ is $A$-harmonic.

Thus the Dirichlet principle and the Dirichlet problem are the same for variational integrals (3.4) and equations (3.5).

$A$-supersolutions are closely related to $A$-*superextremals*; these are defined via an obstacle problem as follows: Let $G$ be a domain in $\mathbf{R}^n$, $h \in W_n^1(G)$, and let $\psi$ be a function defined in $G$. The function $h$ is a "boundary function" and $\psi$ is an "obstacle". Suppose that the family

$$\mathcal{F}_{h,\psi} = \{w \colon w - h \in W_{n,0}^1(G),\ w \geq \psi \text{ a.e.}\}$$

is non-empty. Then we have (see [**GLM1**], [**HKM2**]):

(a) There is a unique $v \in W_n^1(G)$ such that

$$I_F(v) = \inf_{w \in \mathcal{F}_{h,\psi}} I_F(w).$$

(b) The function $v$ is an $A$-supersolution in $G$.

(c) If, in addition, $\psi$ is, in addition, continuous, then $v$ is continuous.

(d) If $v \in W_n^1(G)$ is an $A$-supersolution, $A = \nabla_h F(x, h)$, then $v$ is an $A$-superextremal with boundary values $v$ and $v$ as an obstacle.

Properties (a)–(d) provide the basic steps in the so called obstacle method, which gives a constructive approach to the theory of $A$-superharmonic functions. This method can be extended to the situation where equation (2.6) is not the Euler equation of a variational integral.

# 4. Quasiregular mappings

Let $G$ be a domain in $\mathbf{R}^n$. A continuous mapping $f\colon G \to \mathbf{R}^n$ is said to be $K$-*quasiregular*, abbreviated $QR$, if the component functions of $f$ belong to loc $W_n^1(G)$ and if

$$(4.1) \qquad\qquad |f'(x)|^n \le K\, J(x,f)$$

a.e. in $G$. Here $|f'(x)|$ denotes the supremum norm of the (formal) derivative of $f$, i.e.

$$|f'(x)| = \sup_{|h|=1} |f'(x)\,h|,$$

and $J(x,f)$ is the Jacobian determinant of $f$ at $x$. The term quasiregular is not universally accepted – *bounded distortion* is used by the Soviet school as a synonym. A homeomorphic $QR$ mapping is called *quasiconformal* $(QC)$.

**4.2. Remarks.** (a) For $n = 2$ a mapping $f$ is $1 - QR$ if and only if $f$ is analytic. A $1 - QC$ mapping is conformal. A planar $QR$ mapping $f$ can always be represented in the form $f = g \circ h$ where $g$ is analytic and $h$ is $QC$. Such a representation is not possible in $\mathbf{R}^n$, $n \ge 3$.

(b) For $n \ge 3$ a mapping $f\colon G \to \mathbf{R}^n$ is $1 - QR$ if and only if $f$ is either constant or the restriction of a Möbius transformation of $\mathbf{R}^n \cup \{\infty\}$ to $G$. For this deep result see [**R**].

(c) Although the class of $1 - QR$ mappings is rather narrow in $\mathbf{R}^n$, $n \ge 3$, many interesting examples of $QR$ mappings are now known. The simplest non-injective $QR$ mapping $f\colon \mathbf{R}^3 \to \mathbf{R}^3$ is $(r,\varphi,z) \mapsto (r,2\varphi,z)$ in cylinder coordinates of $\mathbf{R}^3$; $f$ is $4 - QR$.

Partial differential equations provide an important tool for proving the basic properties, (a)–(d) below, of $QR$ mappings. Since this approach is well recorded in Reshetnyak's book [**R**] (see also [**BI**]), we do not explore this difficult subject here.

Let $f\colon G \to \mathbf{R}^n$ be a non-constant $QR$ mapping. We let $B_f$ denote the *branch set* of $f$, i.e. the set of points in $G$ where $f$ does not define a local homeomorphism.

(a) $f$ is a.e. differentiable.

(b) $f$ satisfies Lusin's condition (N) and $J(x,f) > 0$ a.e.

(c) $m(B_f) = 0$.

(d) $f$ is a discrete, open, and sense-preserving mapping.

We recall that $f$ is discrete if $f^{-1}(y)$ is a discrete set of points in $G$ for all $y \in \mathbf{R}^n$, $f$ is open if $f(A)$ is open for each open $A \subset G$, and $f$ is sense-preserving if $\mu(y,f,D) > 0$ for all $y \in f(D) \setminus f(\partial D)$ and all domains $D \Subset G$. Here $\mu$ is the topological degree of the triple $(y,f,D)$; see [**R**]. If $f$ is analytic, then $\mu(y,f,D)$ is sometimes called the winding number of $\partial D$ about $x$, $y = f(x)$. This has a familiar expression as a Cauchy type integral:

$$\mu(y,f,D) = \frac{1}{2\pi i} \int_{\partial D} \frac{dz}{f(z) - y}.$$

The important concepts of normal domain and normal neighborhood in the theory of discrete and open mappings are also very useful: A domain $D \Subset G$ is a *normal domain* of $f$ if $f(\partial D) = \partial f(D)$, and a normal domain $D$ is a *normal neighborhood* of $x$ if $f^{-1}\big(f(x)\big) \cap D = \{x\}$. There is an abundance of normal domains and normal neighborhoods for a non-constant $QR$ mapping $f$: If $G' \subset \mathbf{R}^n$ is an open set and if $D$ is a component of $f^{-1}(G')$ such that $D \Subset G$, then $D$ is a normal domain of $f$. For all sufficiently small $r > 0$ the $x$-component of $f^{-1}(B(f(x), r))$ is a normal neighborhood of $x \in G$. For these properties see [MRV1] or [Ri2].

The use of a normal domain $D$ is based on the multiplicity formula

$$(4.3) \qquad \mu(f(x), f, D) = \sum_{z \in f^{-1}(f(x)) \cap D} i(z, f)$$

for all $x \in D$. Here $i(x, f)$ denotes the local topological degree of $f$ at $x$; for small neighborhoods $\mathcal{U}$ of $x$, $i(x, f) = \mu(f(x), f, \mathcal{U})$.

The most successful analytic tools for studying $QR$ mappings seem to be various capacity and modulus inequalities. These tools are well known in the theory of $QC$ mappings but they also have their counterparts for general $QR$ mappings. Capacity inequalities are related in a natural way to the variational integrals introduced in Section 3. For the use of capacity and modulus inequalities see [MRV1], [Ri2], [V1], [Vu].

## 5. Equations, variational integrals, and $QR$ mappings

It is a fundamental fact in Complex Analysis that if $f : G \to \mathbf{R}^2$ is analytic and $u$ is harmonic, then $u \circ f$ is harmonic. This fact can also be expressed in terms of the Dirichlet principle. Here we study the corresponding properties of $QR$ mappings and $A$-harmonic functions.

Suppose that $A : \mathbf{R}^n \times \mathbf{R}^n \to \mathbf{R}^n$ is a mapping satisfying (2.1)–(2.5). Let $f : G \to \mathbf{R}^n$ be $QR$. Define the *pull back* $f^\# A$ of $A$ as follows

$$f^\# A(x, h) = J(x, f) f'(x)^{-1} A(f(x), f'(x)^{-1^*} h)$$

if $J(x, f) \neq 0$ and $f^\# A(x, h) = A(x, h)$ if $J(x, f) = 0$ or $f'(x)$ does not exist or $x \notin G$. Here $f'(x)^{-1}$ denotes the inverse of the linear mapping $f'(x)$ and $f'(x)^{-1^*}$ is the transpose of $f'(x)^{-1}$. Then $f^\# A : \mathbf{R}^n \times \mathbf{R}^n \to \mathbf{R}^n$ and the proof of the next lemma is not difficult.

**5.1. Lemma.** [MVa], [HKM2] *The mapping $f^\# A$ satisfies the assumptions (2.1)–(2.5).*

**5.2. Remark.** If $f : G \to \mathbf{R}^2$ is analytic or if $f : G \to \mathbf{R}^n$ is a Möbius transformation, then $f^\# A = A$ for the $n$-harmonic operator

$$A(x, h) = |h|^{n-2} h.$$

The invariance of harmonic functions under analytic functions takes the following form for $QR$ mappings.

**5.3. Theorem.** *If $u: G' \to \mathbf{R}$ is an A-harmonic function, then $u \circ f$ is $f^{\#}A$-harmonic in $f^{-1}(G')$.*

There exist several proofs of Theorem 5.3 which depend on assumptions on $A$ and properties of $f$. Reshetnyak [R] uses exterior forms and starts with the $n$-harmonic equation $\nabla \cdot (|\nabla u|^{n-2} \nabla u) = 0$. In fact, one first observes that the function $u(x) = \log|x|$ is $A$-harmonic for $A(x, h) = |h|^{n-2} h$ in $\mathbf{R}^n \setminus \{0\}$ and then proves that $v(x) = \log|f(x)|$ is $f^{\#}A$-harmonic in $G \setminus f^{-1}(0)$. This is a crucial step, especially for property (d) of Section 4. For a similar approach see [BI].

For general operators $A$ there is a more geometric approach which uses the properties (a)–(d) of Section 4. This method can be used to prove the invariance property of $A$-superharmonic functions as well. This approach was pioneered in [GLM1] for variational integrals. We outline the basic steps.

If $f$ is constant, then there is nothing to prove. Assume that $f$ is non-constant. To prove that $v = u \circ f$ is a solution of $\nabla \cdot f^{\#}A = 0$ it suffices to show that $v$ is a solution of $\nabla \cdot f^{\#}A = 0$ in a normal neighborhood $U = U(x_0, f, r)$ (the $x_0$-component of $f^{-1}(B(f(x_0), r))$) of a point $x_0 \in f^{-1}(G')$. Now $f(U) = B = B(f(x_0), r)$. First it is not difficult to see that $v \in C(U) \cap W_n^1(U)$ and hence it remains to show that

$$I = \int_U f^{\#}A(x, \nabla v(x)) \cdot \nabla \psi(x) \, dm(x) = 0$$

whenever $\psi \in C_0^{\infty}(U)$. The idea now is to lift the function $\psi$ to $B$. If $f$ is one-to-one in $U$, then this can be done with the aid of $f^{-1}$. In the non-injective case a more complicated formula is needed: Define $\psi^*: B \to \mathbf{R}$ by

$$\psi^*(y) = \sum_{x \in f^{-1}(y) \cap U} i(x, f) \psi(x).$$

Here $i(x, f)$ is the local topological index of $f$ at $x$. Then $\psi^* \in C_0(B)$ and an argument based on path lifting shows that $\psi^* \in W_n^1(B)$. Next one wants a formula for $\nabla \psi^*(y)$. From (b) and (c) in Section 4 it follows that $m(fB_f) = 0$ and hence that there exists a sequence $B_1, B_2, \ldots$ of disjoint open balls in $V = B \setminus f(U \cap B_f)$ which almost cover $V$ and hence $B$. Now each $B_i$ has exactly $k = i(x_0, f) = \mu(f(x_0), f, D)$ components $U_{ij}$, $j = 1, \ldots, k$, each of which is mapped homeomorphically (quasiconformally) onto $B_i$. Let $f_{ij}$ be the homeomorphisms $f|U_{ij}$ and let $g_{ij} = f_{ij}^{-1}$. If $z \in B_i$, then we have

$$\psi^*(z) = \sum_{j=1}^{k} \psi(g_{ij}(z)),$$

and hence

(5.4)
$$\nabla \psi^*(z) = \sum_{j=1}^{k} f'(g_{ij}(z))^{-1*} \nabla \psi(g_{ij}(z))$$

a.e. in $B_i$. Since $m(B_f) = 0$ and since $\nabla v(x) = f'(x)^* \nabla u(f(x))$ a.e. in $U$, we obtain

$$I = \int_U J(x,f) f'(x)^{-1} A(f(x), \nabla u(f(x))) \cdot \nabla \psi(x) \, dm(x)$$

$$= \sum_{i=1}^{\infty} \sum_{j=1}^{k} \int_{U_{ij}} J(x,f) A(f(x), \nabla u(f(x))) \cdot f'(x)^{-1^*} \nabla \psi(x) \, dm(x)$$

$$= \sum_{i=1}^{\infty} \sum_{j=1}^{k} \int_{B_i} A(y, \nabla u(y)) \cdot f'(g_{ij}(y))^{-1^*} \nabla \psi(g_{ij}(y)) \, dm(y)$$

by the transformation formula for integrals. By (5.4) this yields

$$I = \sum_{i=1}^{\infty} \int_{B_i} A(y, \nabla u(y)) \cdot \nabla \psi^*(y) \, dm(y)$$

$$= \int_B A(y, \nabla u(y)) \cdot \nabla \psi^*(y) \, dm(y).$$

Since $u$ is a solution of $\nabla \cdot A = 0$ and since $\psi^* \in W_{n,0}^1(B)$, the last integral vanishes. Hence $I = 0$, as required.

By the same method one may also show that if $u$ is an $A$-supersolution in $G'$, then $u \circ f$ is an $f^\# A$-supersolution in $f^{-1}(G')$. From property (c) in Section 2 we obtain

**5.5. Theorem.** *If $f: G \to \mathbf{R}^n$ is a non-constant $QR$ mapping and if $u: G' \to \mathbf{R} \cup \{\infty\}$ is $A$-superharmonic, then $u \circ f$ is $f^\# A$-superharmonic.*

There exists another proof for Theorem 5.5 that is potential-theoretic: If $f$ is an $(A^*, A)$-harmonic morphism, then $f$ also preserves superharmonic functions. See the next section.

**5.6. Variational integrals and $QR$ mappings.** Theorem 5.3 has a counterpart for variational integrals. Suppose that a variational kernel $F: \mathbf{R}^n \times \mathbf{R}^n \to \mathbf{R}$ satisfies assumptions (3.1)–(3.4) and let $f: G \to \mathbf{R}^n$ be $QR$. Define the *pull back* $f^\# F$ of $F$ as follows:

$$f^\# F(x, h) = F(f(x), J(x,f)^{1/n} f'(x)^{-1^*} h)$$

if $J(x,f) \neq 0$ and $f^\# F(x,h) = F(x,h)$ if $J(x,f) = 0$ or $f'(x)$ does not exist or $x \notin G$. Then $f^\# F$ satisfies the same assumptions as $F$; the constants $\alpha'$ and $\beta'$ for $f^\# F$ in (3.1) depend on the constants for $F$, on the coefficient $K$ of quasiregularity of $f$, and on $n$; see [GLM1].

A modification of the method used in the proof of Theorem 5.3 gives

**5.7. Theorem.** [GLM1] *If $u$ is a free extremal for $F$ in $G'$, then $u \circ f$ is a free extremal for $f^\# F$ in $f^{-1}(G')$.*

**5.8. Remark.** Extremals and superextremals with boundary values are not, in general, preserved under $QR$ mappings. This is due to the fact that $u \circ f$ need not belong to $W_n^1(f^{-1}(G'))$, although $u \circ f$ is in the corresponding local space. If $u \circ f \in W_n^1(f^{-1}(G'))$, then it is possible to formulate the corresponding results for extremals and superextremals with boundary values and with an obstacle.

## 6. $n$-harmonic morphisms

Let $A$ and $A^*$ be two mappings which satisfy (2.1)–(2.5) with possibly different $\alpha$ and $\beta$. Let $G$ be a domain in $\mathbf{R}^n$. A continuous mapping $f\colon G \to \mathbf{R}^n$ is an $(A^*, A)$-*harmonic morphism* if $u \circ f$ is $A^*$-harmonic in $f^{-1}(G')$ whenever $u$ is $A$-harmonic in $G'$. Further, $f$ is an $n$-*harmonic morphism* if $f$ is an $(A^*, A)$-harmonic morphism for some $A^*$ and $A$.

Theorem 5.3 says that if $f\colon G \to \mathbf{R}^n$ is $QR$, then $f$ is an $(f^\# A, A)$-harmonic morphism for any mapping $A$ satisfying (2.1)–(2.5). Next we turn our attention to the inverse problem. Given an $n$-harmonic morphism $f$ what can be said about $f$? If $n = 2$ and if $A(x, h) = h$ is the Laplace operator, then it is a classical result that an $(A, A)$-harmonic morphism is either analytic or anti-analytic.

**6.1. Theorem.** [**HKM3**] *If $f\colon G \to \mathbf{R}^n$ is a sense-preserving $n$-harmonic morphism, then $f$ is $QR$.*

Note that a sense-preserving mapping in the plane cannot be anti-analytic. It is not known if Theorem 6.1 holds without the condition that $f$ is sense-preserving for $n \geq 3$. In the plane this condition is not needed.

**6.2. Theorem.** [**HKM3**] *If $f\colon G \to \mathbf{R}^2$ is a 2-harmonic morphism, then either $f$ or $f$ composed with a sense-reversing reflection is a quasiregular mapping. In particular, $f$ is of the form $\varphi \circ h$, where $h$ is a quasiconformal homeomorphism and $\varphi$ is either analytic or anti-analytic.*

The proof of Theorem 6.1 is based on a careful analysis of general $n$-harmonic morphisms $f$. If $f$ is not constant, then one first shows that $f$ is open and light. By a light mapping $f$ we mean a mapping such that $f^{-1}(y)$ is totally disconnected for each $y \in \mathbf{R}^n$. This is a considerably weaker requirement than discreteness. However, for $n = 2$ it is known (Stoilow's theorem) that an open and light mapping is discrete. This observation can be used to obtain Theorem 6.2 from Theorem 6.1.

To complete the proof of Theorem 6.1 the metric definition for quasiregularity is employed. A continuous, non-constant mapping $f\colon G \to \mathbf{R}^n$ is $QR$ if and only if the following three conditions hold:

(i) $f$ is sense-preserving, discrete, and open;
(ii) the local distortion $H(x, f)$ is locally bounded in $G$;
(iii) there is a real number $a$ such that $H(x, f) \leq a$ for a.e. $x \in G \setminus B_f$.

Here

$$H(x, f) = \limsup_{r \to 0} \frac{L(x, r, f)}{\ell(x, r, f)}$$

with

$$L(x, r, f) = \sup_{|y-x|=r} |f(y) - f(x)|, \quad \ell(x, r, f) = \inf_{|y-x|=r} |f(y) - f(x)|,$$

and $B_f$, the branch set of $f$, is the set where $f$ fails to be a local homeomorphism. For this characterization see [**MRV1**]. In order to prove (ii) and (iii) for a non-constant $(A^*, A)$-harmonic morphism $f$, one must analyze the singular $A$-harmonic function. The singular $A$-harmonic function is similar to the plane function $\log(1/|x|)$, which is harmonic in $\mathbf{R}^2 \setminus \{0\}$. For the construction of such an $A$-harmonic function $u$ for the general operator $A$ see [HKM2]. The basic method is to compose $u$ with $f$ and study the level lines of $u \circ f$; this is an $A^*$-harmonic function in $G \setminus f^{-1}(0)$. This leads to (ii). An argument based on the use of an $A$-harmonic measure (see Section 7) then gives (iii).

# 7. Applications

As mentioned before, partial differential equations play an essential role in obtaining basic properties (a)–(d) of $QR$ mappings in Section 4. Here we direct our attention to the applications of $A$-harmonic measures. The use of the classical harmonic measure is well known in the theory of conformal and analytic functions.

Suppose that $\Omega$ is an open set, $E \subset \partial\Omega$, and $A \colon \mathbf{R}^n \times \mathbf{R}^n \to \mathbf{R}^n$ satisfies (2.1)–(2.5). The $A$-harmonic measure of $E$ with respect to $\Omega$ at $x \in \Omega$ is defined as

$$(7.1) \qquad \omega(x) = \omega(E, \Omega; A)(x) = \inf\{v(x)\}$$

where the infimum is taken over all non-negative $A$-superharmonic functions $v$ in $\Omega$ such that

$$\liminf_{y \to z} v(y) \geq 1$$

for all $z \in E$. Such a function $v$ is called admissible for $\omega$.

**7.2. Remarks.** (a) If $A(x, h) = h$ is the ordinary Laplace operator, then $\omega$ is the classical (outer) harmonic measure of $E$ with respect to $\Omega$. Definition (7.1) is completely analogous to the classical case.

(b) It can be shown that $\omega$ is $A$-harmonic in $\Omega$ and that $0 \leq \omega \leq 1$; see [GLM2], [HKM2]. In general, for a fixed $x \in G$, $E \mapsto \omega(E, G; A)(x)$ does not define a measure on the Borel subsets $E$ of $\partial\Omega$; this is because of the non-linearity of the equation $\nabla \cdot A = 0$.

With a proper interpretation the $A$-harmonic measure is increased under $QR$ mappings. For such an interpretation fix a mapping $A$ and let $f \colon G \to \mathbf{R}^n$ be $QR$ and non-constant. For $E \subset \partial G$ let $C(E, f)$ denote the cluster set of $E$ under $f$, i.e. $y \in C(E, f)$ if there is a sequence of points $x_i \in G$ such that $x_i \to E$ and $f(x_i) \to y$. The cluster set $C(E, f)$ may contain $\infty$; the definition of $\omega(E, \Omega; A)$ can easily be extended to the case where $E$ contains $\infty$.

The next theorem is a very general form of the principle of harmonic measure for $QR$ mappings. Note that since $f$ is non-constant, $f(G)$ is open and that $C(E, f)$ is always closed.

**7.3. Theorem.** *Under the above assumptions*

$$(7.4) \qquad \omega(E, G; f^\# A)(x) \leq \omega\bigl(C(E, f), f(G) \setminus C(E, f); A\bigr)(f(x))$$

*whenever $x \in G$ and $f(x) \notin C(E, f)$.*

For the proof note that $C(E, f)$ need not be a subset of the boundary of $\Omega = f(G) \setminus C(E, f)$ – the right hand side is actually $\omega^* = \omega(E', \Omega; A)$ where $E' = C(E, f) \cap \partial\Omega$. We shall use this obvious abbreviation. Write $\omega = \omega(E, G; f^\# A)$. To show that $\omega(x) \leq$

$\omega^*\big(f(x)\big)$ let $u$ be an $A$-superharmonic function admissible for $\omega^*$. By Theorem 5.5, $u \circ f$ is admissible for $\omega_1 = \omega\big(E^*, f^{-1}(\Omega); f^\# A\big)$ where $E^* = E \cup f^{-1}\big(C(E, f)\big)$. Now for each $\varepsilon > 0$, $v_\varepsilon = \min(1, u \circ f + \varepsilon)$ is $f^\# A$-superharmonic in $G$ and admissible for $\omega$. Thus $\omega(x) \le v_\varepsilon(x) \le u\big(f(x)\big) + \varepsilon$ for each $x \in G$ with $f(x) \in \Omega$, and since $u$ and $\varepsilon$ were arbitrary we obtain $\omega(x) \le \omega^*\big(f(x)\big)$, as required.

Suppose that $f$ is $QC$. Then $f^{-1\#} f^\# A = A$, and hence Theorem 7.3 yields

**7.5. Corollary.** *Suppose that $f: \bar{G} \to \bar{G}'$ is a homeomorphism and that $f|G$ is $QC$. Then*

$$(7.6) \qquad \omega(E, G; f^\# A)(x) = \omega\big(f(E), f(G); A\big)\big(f(x)\big)$$

*for every $E \subset \partial G$ and for every $x \in G$.*

For $n = 2$ and $A(x, h) = h$ it is a famous result due to Beurling and Ahlfors [**BA**] that there is no relation between the left and right hand sides of (7.6) if $f^\# A$ is replaced by $A$; see also [**CFK**]. The corresponding question is open in $\mathbf{R}^n$, $n \ge 3$.

Since an $A$-harmonic measure is not a measure in general, $A$-harmonic measures $\omega$ have been used mostly in the situations where $\omega$ can be estimated in terms of the geometry of $E$. Applications to $QR$ mappings include
  (a) two-constant theorem [**Ri1**], [**GLM2**],
  (b) Phragmen–Lindelöf-type results [**GLM3**],
  (c) Lindelöf's theorem on asymptotic limits [**GLM3**], [**Ri1**], [**Vu**],
  (d) Milloux's problem [**M1**], and
  (e) Øksendal's theorem for $QC$ mappings [**HM**], [**He**], [**M3**].
An interesting problem, also related to the aforementioned applications, is to determine sets of $A$-harmonic measure zero. In general, the class of sets $E \subset \partial\Omega$ such that $\omega(E, \Omega; A) = 0$ depends on $A$. However, there is a stable subclass: A set $E \subset \partial\Omega$ is said to be of *total $n$-harmonic measure zero* if $\omega(E, \Omega; A) = 0$ for all $A$ satisfying (2.1)–(2.5). There is an abundance of sets of total $n$-harmonic measure zero. For simplicity consider the unit ball $B$ in $\mathbf{R}^n$. Let $E \subset \partial B$ and $r > 0$. Let $E_r$ denote the $r$-inflation of $E$, i.e.

$$E_r = \bigcup_{x \in E} B(x, r),$$

and write $\delta(r) = \sup_{y \in E} \operatorname{dist}(y, \partial B \setminus E_r)$.

**7.7. Theorem.** [**M5**] *If $\liminf_{r \to 0} \delta(r)/r < \infty$, then $E$ is of total $n$-harmonic measure zero.*

Theorem 7.7 implies, for example, that a smooth curve on $\partial B$ in $\mathbf{R}^n$, $n \ge 3$, is of total harmonic measure zero. For $n = 2$ such a curve has always positive $A$-harmonic measure. For this reason the boundary behavior of $QR$ mappings in $B$ is quite different for $n = 2$ and for $n \ge 3$. A typical example is application (c) above. Let $\gamma$ be a curve in the unit ball $B$ of $\mathbf{R}^n$ with one endpoint $b$ in $\partial B$. For $n = 2$ the classical theorem of Lindelöf says that if $f: B \to \mathbf{R}^2$ is a bounded analytic function and if $f(x)$ tends to a limit as $x$ tends to $b$ along $\gamma$, then $f$ has the same limit in every Stolz angle at $b$. This statement holds for plane $QR$ mappings as well. Now for $n \ge 3$ this is not true; see

[**Ri1**]. However, there is a version of this result in $\mathbf{R}^n$, $n \geq 2$. Let $L$ be the $x_1$-axis in $\mathbf{R}^n$ and let $\gamma$ be a curve in $B \setminus L$ ending at 0. If $f : B \setminus L \to \mathbf{R}^n$ is a bounded $QR$ mapping and if $f(x)$ tends to a limit as $x$ tends to 0 along $\gamma$, then $f(x)$ tends to the same limit in each Stolz "cone" at 0 in $B \setminus L$; for $n = 2$, $B \setminus L$ is disconnected and the Stolz angle must lie in the same half-plane as $\gamma$. The reason for this result is that $L$ and $\gamma$ are of the same size from the $A$-harmonic measure-theoretic point of view; see [**GLM3**].

Note also that sets of total harmonic measure zero cannot be characterized in terms of the Hausdorff dimension. For $n = 2$ there are sets $E$ on the boundary of the unit disk $B$ whose Hausdorff dimension is arbitrary near 0 but $E$ is not of total harmonic measure zero; see [**Tu**], [**M4**]. On the other hand, it follows from Theorem 7.7 that $\frac{1}{3}$-Cantor set constructed on $\partial B$ is of total $n$-harmonic measure zero.

# References

[BJS]     Bers, L., F. John, and M. Schechter, Partial Differential Equations, Interscience Publishers, New York, 1964.

[BA]      Beurling, A., and L. Ahlfors, The boundary correspondence under quasiconformal mappings, Acta Math. 96 (1956), 125–142.

[BI]      Bojarski, B., and T. Iwaniec, Analytic foundations of the theory of quasiconformal mappings in $\mathbf{R}^n$, Ann. Acad. Sci. Fenn. Ser. A I Math. 8 (1983), 234–257.

[CFK]     Caffarelli, L., E. Fabes, and C. Kenig, Completely singular elliptic harmonic measures, Indiana Univ. Math. J. 30 (1981), 917–924.

[G]       Gehring, F. W., The $L^p$-integrability of the partial derivatives of a quasiconformal mapping, Acta Math. 130 (1973), 265–277.

[GLM1]    Granlund, S., P. Lindqvist, and O. Martio, Conformally invariant variational integrals, Trans. Amer. Math. Soc. 277 (1983), 43–47.

[GLM2]    Granlund, S., P. Lindqvist, and O. Martio, $F$-harmonic measure in space, Ann. Acad. Sci. Fenn. Ser. A I Math. 7 (1982), 233–247.

[GLM3]    Granlund, S., P. Lindqvist, and O. Martio, Phragmén–Lindelöf's and Lindelöf's theorems, Ark. Mat. 23 (1985), 103–128.

[He]      Heinonen, J., Boundary accessibility and elliptic harmonic measures, Complex Variables 10 (1988), 273–282.

[HK1]     Heinonen, J., and T. Kilpeläinen, $A$-superharmonic functions and supersolutions of degenerate elliptic equations, Ark. Mat. 26 (1988), 87–105.

[HK2]     Heinonen, J., and T. Kilpeläinen, Polar sets for supersolutions of degenerate elliptic equations, Math. Scand. (to appear).

[HM]      Heinonen, J., and O. Martio, Estimates for $F$-harmonic measures and Øksendal's theorem for quasiconformal mappings, Indiana Univ. Math. J. 38 (1987), 659–683.

[HKM1]    Heinonen, J., T. Kilpeläinen, and O. Martio, Fine topology and quasilinear elliptic equations, Ann. Inst. Fourier (Grenoble) 39, 2 (1989), 293–318.

[HKM2]    Heinonen, J., T. Kilpeläinen, and O. Martio, Non-linear Potential Theory (to appear).

[HKM3]    Heinonen, J., T. Kilpeläinen, and O. Martio, Harmonic morphisms for quasilinear elliptic partial differential equations. (to appear).

[I]       Iwaniec, T., Some aspects of partial differential equations and quasiregular mappings, Proceedings of ICM, Warsaw 1983, Warsaw, 1985, pp. 1193–1208.

[LV]      Lehto, O., and K. I. Virtanen, Quasikonforme Abbildungen, Springer-Verlag, 1965.

[LM1]     Lindqvist, P., and O. Martio, Two theorems of N. Wiener for solutions of quasilinear elliptic equations, Acta Math. 155 (1985), 153–171.

[LM2]     Lindqvist, P., and O. Martio, Regularity and polar sets for supersolutions of certain degenerate elliptic equations, J. Analyse Math. 50 (1988), 1–17.

[M1]      Martio, O., $F$-harmonic measures, quasihyperbolic distance and Milloux's problem, Ann. Acad. Sci. Fenn. Ser. A I Math. 12 (1987), 151–162.

[M2]      Martio, O., Counterexamples for unique continuation, Manuscripta Math. 60 (1988), 21–47.

[M3]    Martio, O., Sets of zero elliptic harmonic measures, Ann. Acad. Sci. Fenn. Ser. A I Math. 14 (1989), 47–55.

[M4]    Martio, O., Potential theoretic aspects of non-linear elliptic partial differential equations, Dept. of Math., Univ. of Jyväskylä, Report 44 (1989), 1–23.

[M5]    Martio, O., Harmonic measures for second order non-linear partial differential equations, Function spaces, differential operators and non-linear analysis, L. Päivärinta, ed., Pitman Research Notes in Mathematics, 1989, pp. 271–279.

[MRV1]  Martio, O., S. Rickman, and J. Väisälä, Definitions for quasiregular mappings, Ann. Acad. Sci. Fenn. Ser. A I Math. 448 (1969), 1–40.

[MRV2]  Martio, O., S. Rickman, and J. Väisälä, Distortion and singularities of quasiregular mappings, Ann. Acad. Sci. Fenn. Ser. A I Math. 465 (1970), 1–13.

[MVa]   Martio, O., and J. Väisälä, Elliptic equations and maps of bounded length distortion, Math. Ann. 282 (1988), 423–443.

[Maz]   Maz'ya, V. G., On the continuity at a boundary point of solutions of quasi-linear elliptic equations, Vestnik Leningrad Univ. 3 (1976), 225–242. (English translation).

[Mo]    Morrey, C. B., On the solutions of quasilinear elliptic partial differential equations, Trans. Amer. Math. Soc. 43 (1938), 126–166.

[Ra]    Radó, T., Subharmonic Functions, Ergebnisse der Mathematik und Ihrer Grenzgebiete, Chelsea Publishing Company, 1949.

[R]     Reshetnyak, Yu. G., Space Mappings with Bounded Distortion, AMS Trans. of Math. Monographs vol. 73, 1989.

[Ri1]   Rickman, S., Asymptotic values and angular limits of quasiregular mappings of a ball, Ann. Acad. Sci. Fenn. Ser. A I Math. 5 (1980), 185–196.

[Ri2]   Rickman, S., Quasiregular mappings. (to appear).

[S]     Serrin, J., Local behavior of solutions of quasi-linear equations, Acta Math. 111 (1964), 247–302.

[T]     Tsuji, M., Potential Theory in Modern Function Theory, Maruzen Co., Tokyo, 1959.

[Tu]    Tukia, P., Hausdorff dimension and quasisymmetric mappings, Math. Scand. 65 (1989), 152–160.

[V1]    Väisälä, J., Lectures on $n$-dimensional Quasiconformal Mappings, Lecture Notes in Mathematics 229, Springer-Verlag, 1971.

[V2]    Väisälä, J., A survey of quasiregular maps in $\mathbf{R}^n$, Proceedings of ICM, Helsinki 1978, vol. 2, Helsinki, 1980, pp. 685–691.

[Ve]    Vekua, N., Generalized Analytic Functions, Pure and Applied Math. 25, Pergamon Press, 1962.

[Vu]    Vuorinen, M., Conformal Geometry and Quasiregular Mappings, Lecture Notes in Mathematics 1319, Springer-Verlag, 1988.

Quasiconformal Space Mappings
– A collection of surveys 1960–1990
Springer–Verlag (1992), 80–92
Lecture Notes in Mathematics Vol. 1508

# ON FUNCTIONAL CLASSES INVARIANT
# RELATIVE TO HOMOTHETIES

Yu. G. Reshetnyak

Institute of Mathematics, 630090 Novosibirsk, USSR

## 1. Introduction

A. P. Kopylov [1] made an interesting observation consisting of the following. Suppose that we have a functional class whose elements are functions defined on open subsets of $\mathbf{R}^n$ and which take their values in $\mathbf{R}^m$ . Suppose that every subset of this class which consists of functions defined on the same open $U \subset \mathbf{R}^n$, and is bounded in the sense of the uniform norm, is uniformly equicontinuous on every compact subset of $U$. If the class considered is also invariant in a natural sense relative to homotheties in $\mathbf{R}^n$ and in $\mathbf{R}^m$ then every function belonging to this class will satisfy a Hölder condition on every compact subset of its domain of definition.

The main aim of this paper is to present a simplified proof of this result of A. P. Kopylov. We shall also give here some applications of this result. In fact, A. P. Kopylov proved a more general assertion which will also be considered.

In what follows, $\mathbf{R}^n$ will denote the $n$–dimensional euclidean space of vectors $x = (x_1, \ldots, x_n)$, where $x_1, x_2, \ldots, x_n$ are real numbers. For vectors $x = (x_1, x_2, \ldots, x_n)$, $y = (y_1, y_2, \ldots, y_n)$ let $\langle x, y \rangle = x_1 y_1 + x_2 y_2 + \ldots + x_n y_n$ be the scalar product of $x$ and $y$. For $x \in \mathbf{R}^n$, $|x| = (\langle x, x \rangle)^{1/2}$ is the euclidean norm of $x$. The mapping $\alpha : \mathbf{R}^n \to \mathbf{R}^n$ is said to be a homothety if for every $x \in \mathbf{R}^n$, $\alpha(x) = \lambda x + a$, where $\lambda > 0$, $\lambda \in \mathbf{R}$, $a \in \mathbf{R}^n$ are constants.

For a point $a \in \mathbf{R}^n$ and a real number $r > 0$ let $B(a,r) = \{x \in \mathbf{R}^n : |x - a| < r\}$, $\overline{B}(a,r) = \{x \in \mathbf{R}^n : |x - a| \leq r\}$; $\overline{B}(a,r)$ is a closed ball, $B(a,r)$ is an open ball in $\mathbf{R}^n$; $a$ is the center, $r$ is the radius of both these balls.

For an arbitrary nonempty set $E \subset \mathbf{R}^n$, $\Delta(E)$ is the diameter of $E$, i.e.

$$\Delta(E) = \sup_{x,y \in E} |x - y|.$$

Let $U$ be an open set in $\mathbf{R}^n$. We shall say that the set $A$ is strictly inside $U$ if the closure $\overline{A}$ of the set $A$ is compact and $U \supset \overline{A}$.

Let $U$ be an arbitrary open subset of the space $\mathbf{R}^n$. For a mapping $f : U \to \mathbf{R}^m$ and a set $E \subset U$ let

$$\Delta(f, E) = \Delta(f(E)), \quad \|f\|_{L_\infty(E)} = \sup_{x \in E} |f(x)|.$$

For a measurable set $E \subset \mathbf{R}^n$, meas $E$ will denote the $n$–dimensional Lebesgue measure of $E$.

We shall consider the classes of mappings $f : U \to \mathbf{R}^m$, where $U$ is an arbitrary (variable) open subset of $\mathbf{R}^n$. The class of mappings $\mathfrak{G}$ is said to be of type $K(n,m)$ if the following conditions K1, K2, and K3 hold.

K1. If the mapping $f : U \to \mathbf{R}^n$ belongs to $\mathfrak{G}$, where $U$ is an open subset of $\mathbf{R}^n$, then for every pair of homotheties $\alpha : \mathbf{R}^m \to \mathbf{R}^m$, $\beta : \mathbf{R}^n \to \mathbf{R}^n$ acting in $\mathbf{R}^m$ and $\mathbf{R}^n$ the composition

$$\alpha \circ f \circ \beta$$

also belongs to the class $\mathfrak{G}$.

K2. Let $f : U \to \mathbf{R}^m$ be an arbitrary mapping of the class $\mathfrak{G}$. Then for every open $V \subset U$ the restriction $f|V$ of $f$ to $V$ is also an element of the class $\mathfrak{G}$.

K3. Let $\mathfrak{G}_0$ be the set of all mappings of the class $\mathfrak{G}$ which are defined on the unit ball $B(0,1)$ and which satisfy the following conditions: $f(0) = 0$ and $|f(x)| \leq 1$ for all $x \in B(0,1)$. Then there exists a constant $d \in (0,1)$ such that the set $\mathfrak{G}_0$ is uniformly equicontinuous on the closed ball $\overline{B}(0,d)$.

**Theorem 1.** *Let $\mathfrak{G}$ be a class of mappings of type $K(n,m)$. Then there exists a constant $\alpha \in (0,1)$, such that for every ball $B(a,r)$ in $\mathbf{R}^n$ and for every bounded function $f : B(a,r) \to \mathbf{R}^m$ belonging to the class $\mathfrak{G}$*

$$|f(x) - f(a)| \leq 2\Delta(f, B(a,r))(|x - a|/r)^\alpha$$

*for all $x \in B(a,r)$.*

**Corollary.** *Let $\mathfrak{G}$ be a functional class of type $K(n,m)$, and $U$ an open set in $\mathbf{R}^n$. Then for every compact $A \subset U$ there exists a finite $C \geq 0$ such that*

$$|f(x_1) - f(x_2)| \leq C\|f\|_{L_\infty(U)}|x_1 - x_2|^\alpha$$

*for every bounded mapping $f : U \to \mathbf{R}^m$ belonging to the class $\mathfrak{G}$ and for every pair $x_1, x_2$ of points of the set $A$. Here $\alpha$ is the same as in Theorem 1.*

Theorem 1 and its corollary were first proved in [1] (see also [2]).

Here we shall consider also some applications of Theorem 1 and of its corollary. Specifically, it will be shown how with the help of this theorem one can get a simple proof of the Hölder continuity of extremals of some variational functionals. (This result can be established also in other ways; see for instance [3], [4], [5]). The other application of Theorem 1 is the proof of the Hölder continuity of quasiconformal and quasiregular space mappings. Using the theorem of A. P. Kopylov formulated above one can deduce the Hölder continuity of such mappings from an old result of M.A. Kreines [6]. However, the precise value of the exponent $\alpha$ in the Hölder condition can not be found in this way. We omit any details connected with the application of Theorem 1 to quasiconformal and quasiregular mappings.

In fact, A. P. Kopylov [1] established an assertion which is more general than Theorem 1. This more general result will be considered here, but first we shall introduce some auxiliary concepts which are necessary to formulate the general theorem of A. P. Kopylov.

Let $\mathfrak{G}$ be a functional class of type $K(n,m)$. Let us fix a number $\rho$ from the open interval $(0,1)$. Suppose that $f : U \to \mathbf{R}^m$ is a continuous mapping, where $U$ is an open set in $\mathbf{R}^n$. Let $B = B(x,r)$ be a ball such that $U \supset B$. We set $B_\rho = B(x,\rho r)$. At first we shall define a number $\xi_{\rho,B}(f,\mathfrak{G})$. If $\Delta(f,B)$ is equal either to 0 or to $\infty$ then we set $\xi_{\rho,B}(f,\mathfrak{G}) = 0$. In the case where $0 < \Delta(f,B) < \infty$ let

$$(1) \qquad \xi_{\rho,B}(f,\mathfrak{G}) = \inf_{g \in \mathfrak{G}} \|f - g\|_{L_\infty(B_\rho)}/\Delta(f,B),$$

where the greatest lower bound is taken over the set of all mappings $g$ of the class $\mathfrak{G}$ which are defined on the ball $B$. It is necessary to say here that the $L_\infty$-norm in (1) is taken over the ball $B_\rho = B(x,\rho r)$. Finally, let

$$\xi_\rho(f,\mathfrak{G}) = \sup_{B \subset U} \xi_{\rho,B}(f,\mathfrak{G}),$$

where the least upper bound is taken over the set of all balls $B$ contained in $U$.

**Theorem 2 ([1],[2]).** *Let $\mathfrak{G}$ be a class of mappings of type $K(n,m)$. Then for every $\gamma \in (0,1/2)$ and for every $\rho \in (0,1)$ there exist constants $\alpha \in (0,1)$ and $C \in (0,\infty)$ such that for every bounded mapping $f : B(a,r) \to \mathbf{R}^m$ satisfying the condition*

$$\xi_\rho(f,\mathfrak{G}) \leq \gamma,$$

*the following inequality holds*

$$|f(x) - f(a)| \leq C\Delta(f,B(a,r))(|x - a|/r)^\alpha$$

*for all $x \in B(a,r)$.*

Theorem 2 admits a corollary analogous to that of Theorem 1. We leave this to the reader to formulate.

Formally, Theorem 1 is a consequence of Theorem 2. But we shall give the proof of Theorem 1 separately because in this case the necessary reasoning will be a little simpler, and for applications Theorem 1 probably is more useful than Theorem 2. The formulation of this last theorem uses a certain quite special construction. The proof of Theorem 2, which will be presented here uses the result of Theorem 1, but by a small modification of this proof one can avoid the reference to Theorem 1.

## 2. The proof of Theorem 1

**Lemma 1.** *Let $\omega(t)$, $0 \le t \le 1$, be a nonnegative nondecreasing function. Suppose that there exist numbers $\theta, \eta \in (0,1)$ such that*

$$\omega(\theta t) \le \eta \omega(t)$$

*for every $t \in [0,1]$. Then for every $t \in [0,1]$*

(2.1) $$\omega(t) \le (\omega(1)/\eta)t^\alpha,$$

*where $\alpha = (\ln \eta)/\ln \theta > 0$.*

**Proof.** Let us take an arbitrary $t \in [0,1]$. In the cases $t = 0$, $t = 1$, the inequality (2.1) evidently is true. Suppose that $0 < t < 1$. Let us find a natural number $m$ such that $\theta^m \le t < \theta^{m-1}$. From the conditions of the lemma it obviously follows that

$$\omega(t) \le \omega(\theta^{m-1}) \le \eta^{m-1}\omega(1) = (\omega(1)/\eta)\,\exp(m \ln\, \eta).$$

Further, $m \ln \theta \le \ln t$ and then $m \ln\, \eta \le (\ln \eta/\ln \theta)\ln t = \alpha \, \ln t$. Hence

$$\omega(t) \le (\omega(1)/\eta)\exp(\alpha \ln t) = (\omega(1)/\eta)t^\alpha.$$

The lemma is proved.

Let $\mathfrak{G}$ be a class of mappings of type $K(n,m)$, and let $\mathfrak{G}_0$ be the set of all functions $f : B(0,1) \to \mathbf{R}^m$ for which $f(0) = 0$ and $|f(x)| \le 1$ for all $x \in B(0,1)$. For a mapping $f \in \mathfrak{G}_0$ and for $t \in [0,1]$ we set

$$\omega_f(t) = \sup_{|x| \le t} |f(x)|.$$

Let $\omega(t)$ be the least upper bound of $\omega_f(t)$ taken over all $f \in \mathfrak{G}_0$. From the definition it is clear that $\omega(t)$ is nondecreasing and $0 \le \omega(t) \le 1$ for all $t \in [0,1]$. According to condition K3, from the definition of classes of type $K(n,m)$, there exists a number $d \in (0,1)$ such that the set of mappings $\mathfrak{G}_0$ is uniformly equicontinuous on the ball $\overline{B}(0,d)$. This fact obviously permits us to conclude that $\omega(t) \to 0$ as $t \to 0$.

We show now that there exists $\theta$ from the open interval $(0,1)$ such that for every $t \in [0,1]$

(2.2) $$\omega(\theta t) \le \frac{\omega(t)}{2}.$$

Let us assume, to the contrary, that no such $\theta$ exists. Let us take an arbitrary natural number $m$. Then, in view of this antithesis, there exists at least one $t = t_m \in [0,1]$ for which

$$\omega(t_m/m) > \frac{1}{2}\omega(t_m).$$

Indeed, if there is no such $t \in [0,1]$ then for every $t \in [0,1]$ we shall have $\omega(t/m) \le \frac{1}{2}\omega(t)$ and we see that in this case the inequality (2.2) will be true with $\theta = 1/m$.

According to the definition of $\omega(t)$ there exists a function $f \in \mathfrak{G}_0$ such that

$$\omega_f(t_m/m) > \frac{1}{2}\omega(t_m) \geq \frac{1}{2}\omega_f(t_m).$$

From the continuity of the function $f$ it follows that there exists a point $x_m$ such that $|x_m| \leq t_m/m$ and $|f(x_m)| = \omega_f(t_m/m)$. Let

$$F(y) = \frac{1}{\omega_f(t_m)}f(t_m y), \quad y_m = \frac{1}{t_m}x_m.$$

Obviously we have $F(0) = 0$ and $|F(y)| \leq 1$ for all $y \in B(0,1)$. From condition K1 in the definition of the functional classes of type $K(n,m)$ it follows that $F \in \mathfrak{G}$. Then $F \in \mathfrak{G}_0$ since $F(0) = 0$ and $\|F\|_{L_\infty(B(0,1))} \leq 1$. Further, we see that $|y_m| = |x_m|/t_m \leq 1/m$. Hence we conclude that $|F(y_m)| \leq \omega(1/m)$. On the other hand,

$$|F(y_m)| = |f(x_m)|/\omega_f(t_m)$$

$$= \omega_f(t_m/m)/\omega_f(t_m) \geq \frac{1}{2}$$

and we conclude that

$$\omega(1/m) \geq \frac{1}{2}$$

for every natural number $m$. So we obviously come to a contradiction of the fact that $\omega(t) \to 0$ as $t \to 0$. This contradiction shows the existence of a constant $\theta \in (0,1)$ for which $\omega(\theta t) \leq \frac{1}{2}\omega(t)$ for every $t \in [0,1]$. Applying Lemma 1 we conclude that for each $t \in [0,1]$

$$\omega(t) \leq (2\omega(1))t^\alpha = 2t^\alpha,$$

where $\alpha = (\ln\frac{1}{2})/\ln\theta > 0$. According to the definition of $\omega(t)$, for every function $f \in \mathfrak{G}_0$ we have the inequality $|f(x)| \leq \omega(|x|)$, and thus for every $x \in B(0,1)$

(2.3) $$|f(x)| \leq 2|x|^\alpha.$$

Let $B(a,r)$ be an arbitrary ball in $\mathbf{R}^n$ and let $f : B(a,r) \to \mathbf{R}^n$ be a bounded mapping of the class $\mathfrak{G}$. In the case where $f$ is identically constant the inequality of Theorem 1 is obvious. Suppose that $f$ is not identically constant. Then $0 < \Delta(f, B(a,r)) = \Delta_o < \infty$. For $x \in B(0,1)$ let $\varphi(x)$ be equal to $\frac{1}{\Delta_o}[f(a+rx) - f(a)]$. In view of condition K1, in the definition of the functional classes of type $K(n,m)$, the mapping $\varphi$ belongs to $\mathfrak{G}$. Evidently $\varphi(0) = 0$, and $|\varphi(x)| \leq 1$ for all $x \in B(0,1)$; hence $\varphi$ belongs to $\mathfrak{G}_0$. Setting $f = \varphi$ in (2.3), we get $|\varphi(x)| \leq 2|x|^\alpha$ for every $x \in B(0,1)$. Changing $x$ to $(x-a)/r$ in the last inequality, we arrive at

$$|f(x) - f(a)| \leq 2\Delta_o(|x - a|/r)^\alpha$$

and Theorem 1 is proved.

The proof of the corollary is based on some simple arguments using the Borel lemma. We leave all details to the reader.

### 3. The proof of Theorem 2

Let $\mathfrak{G}$ be a functional class of type $K(n, m)$. Fix an arbitrary $\rho \in (0, 1)$. Denote by $\mathfrak{H}_0$ the set of all continuous mappings $f : B(0, 1) \to \mathbf{R}^m$ such that $f(0) = 0$, $\Delta(f, B(0, 1)) \leq 1$, $\xi_\rho(f, \mathfrak{G}) \leq \gamma$. For the sake of simplicity we shall denote by $B$ and $B_r$ the balls $B(0, 1)$ and $B(0, r)$, respectively. For $f \in \mathfrak{H}_0$ let

$$\delta_f(t) = \Delta(f, B_t),$$

where $0 < t \leq 1$. Further, let

$$\delta(t) = \sup_{f \in \mathfrak{H}_0} \delta_f(t).$$

It is clear that $\delta$ is a nondecreasing function in the interval $0 < t \leq 1$ and that $0 \leq \delta(t)$ for all $t \in [0, 1]$. We put $\delta(0) = 0$.

In accordance with the formulation of Theorem 2 suppose that $0 < \gamma < \frac{1}{2}$. Let $\eta = \gamma + \frac{1}{2}$. Then

(3.1) $$0 < \eta < 1, \quad \gamma < \frac{\eta}{2}.$$

We shall prove that there exists a constant $\theta$ such that $0 < \theta < 1$ and such that

(3.2) $$\delta(\theta t) \leq \eta \delta(t)$$

for all $t \in [0, 1]$. Suppose, to the contrary, that no such constant $\theta$ exists. Let us take an arbitrary natural number $m$. Then, in view of this antithesis, there exists at least one $t = t_m \in [0, 1]$ for which

(3.3) $$\delta(\frac{t_m}{m}) > \eta \delta(t_m).$$

Indeed, if for some $m$ there is no such $t = t_m$ then $\delta(t/m) \leq \eta \delta(t)$ for all $t \in [0, 1]$ and we see that inequality (3.2) is satisfied with the constant $\theta = \frac{1}{m}$. Since $\delta(t) \geq 0$ for all $t$, from (3.3) it follows that $\delta(t_m/m) > 0$ for all $m$. Since $t_m/m \to 0$ when $m \to \infty$ and since the function $\delta$ is nondecreasing we conclude that $\delta(t) > 0$ for every $t$ such that $0 < t \leq 1$. By the definition of $\delta(t)$ there exists a mapping $f \in \mathfrak{H}_0$ such that

$$\delta_f(\frac{t_m}{m}) > \eta \delta(t_m) \geq \eta \delta_f(t_m).$$

In particular, we see that $\delta_f(t_m/m) > 0$ for all $m$ and thus $\delta_f(t) > 0$ for every $t \in (0, 1]$. Let

$$\varphi : x \in B(0, 1) \mapsto \frac{1}{\delta_f(t_m)} f(x t_m).$$

Obviously $\xi_\rho(\varphi, \mathfrak{G})$ is equal to $\xi_\rho(\tilde{f}, \mathfrak{G})$, where $\tilde{f}$ is the restriction of the mapping $f$ to the ball $B_{t_m}$. We have

$$\xi_\rho(\tilde{f}, \mathfrak{G}) \leq \xi_\rho(f, \mathfrak{G})$$

and thus

$$\xi_\rho(\varphi, \mathfrak{G}) \leq \xi_\rho(f, \mathfrak{G}) \leq \gamma.$$

Further, $\varphi(0) = 0$ and $|\varphi(x)| \leq 1$ for all $x \in B$. Hence $\varphi \in \mathfrak{H}_0$. It is clear that

$$\delta_\varphi(\frac{1}{m}) = \delta_f(\frac{t_m}{m})/\delta_f(t_m) > \eta > 0.$$

Suppose that $\frac{1}{m} < \frac{\rho}{4}$. According to the definition of $\delta(\frac{1}{m})$ there exist points $p, q \in B_{1/m}$ such that $|\varphi(p) - \varphi(q)| > \eta$. Let $\zeta = \frac{\eta}{4} + \frac{\gamma}{2}$. From (3.1) it obviously follows that

(3.4)
$$\gamma < \zeta < \frac{\eta}{2}.$$

The definition of the quantity $\xi_\rho(f, \mathfrak{G})$ permits us to conclude that

$$\xi_{\rho,B}(f, \mathfrak{G}) \leq \xi_\rho(\varphi, \mathfrak{G}) \leq \gamma,$$

and since $\gamma < \zeta$ there exists a function $g \in \mathfrak{G}$ which is defined in the ball $B$ and satisfies the inequality

$$\frac{\|\varphi - g\|_{L_\infty(B_\rho)}}{\Delta(\varphi, B)} < \zeta.$$

Since $\Delta(\varphi, B) \leq 1$ then

$$\|\varphi - g\|_{L_\infty(B_\rho)} < \zeta$$

and thus $|g(x)| \leq \zeta + |\varphi(z)| \leq \zeta + 1$. We have $|p| \leq \frac{1}{m}$, $|q| \leq \frac{1}{m}$, and thus $|p - q| < \frac{2}{m} \leq \frac{\rho}{2}$. It follows that the ball $B(p, \frac{\rho}{2})$ lies in $B_\rho$. The mapping $g$ belongs to the class $\mathfrak{G}$, which is of type $K(n, m)$, and thus in view of Theorem 1

$$|g(q) - g(p)| \leq 2^{1+\beta}(\zeta + 1)(\frac{|p-q|}{\rho})^\beta$$

for some $\beta > 0$. Consequently

$$|g(q) - g(p)| \leq \frac{C}{m^\alpha},$$

where $C < \infty$ is a constant. Further, we have

$$\eta < |\varphi(p) - \varphi(q)| \leq |\varphi(p) - \varphi(q)| + |g(p) - q(q)| + |g(q) - \varphi(q)| < 2\zeta + \frac{C}{m^\alpha}.$$

Passing to the limit as $m \to \infty$ we get $\eta \leq 2\zeta$, which contradicts (3.4).

It was the assumption that there exists no $\theta \in (0, 1)$ such that inequality (3.2) is true for all $t \in [0, 1]$ that led to this contradiction. Hence the existence of such $\theta$ follows.

From Lemma 1 we conclude that

$$\delta(t) \leq Ct^\alpha,$$

where $\alpha = (\ln \eta)/\ln \theta > 0$, $C = \delta(1)/\eta \leq \frac{1}{\eta}$.

Let $f : B(a, r) \to \mathbf{R}^m$ be an arbitrary bounded mapping such that $\xi_\rho(f, \mathfrak{G}) \le \gamma$. Then the function

$$\varphi(x) = \frac{f(a + rx) - f(a)}{\Delta(f, B(a, r))}$$

belongs to the class $\mathfrak{H}_0$. (It is supposed that $f$ is not identically constant). Consequently,

$$|\varphi(x)| \le C|x|^\alpha.$$

Replacing $x$ here by $x - a/r$ , where $x \in B(a, r)$, we get

$$|f(x) - f(a)| \le C\Delta(f, B(a, r))(|x - a|/r)^\alpha.$$

Theorem 2 is proved.

## 4. Some applications of Theorem 1

First of all let us formulate a lemma about the continuity of mappings of the class $W^1_{n,loc}(U)$.

Let $U$ be an open set $\mathbf{R}^n$. We shall say that a function

$$f : U \to \mathbf{R}^m, \quad f(x) = (f_1(x), f_2(x), \dots, f_m(x))$$

belongs to the class $W^1_{p,loc}(U)$ if each of the real functions $f_1, f_2, \dots, f_m$, which are the components of the vector function $f$, has distributional first partial derivatives that are locally summable of degree $p$ in $U$. For a measurable set $A \subset U$ let

$$\|f\|_{L^1_p(A)} = \sum_{i=1}^m \sum_{j=1}^n \|\frac{\partial f_i}{\partial x_j}\|_{L_p(A)},$$

$$\|f\|_{W^1_p(A)} = \|f\|_{L_p(A)} + \|f\|_{L^1_p(A)}.$$

Let $U$ be an open subset of $\mathbf{R}^n$, and $f : U \to \mathbf{R}^m$ a continuous mapping. We say that $f$ satisfies condition $E(T)$, where $T \ge 1$ is a constant, if

$$\Delta(f, \overline{B}(a, r)) \le T\Delta(f, S(a, r))$$

for every closed ball $\overline{B}(a, r) \subset U$.

**Lemma 2.** Let $U$ be an open set of $\mathbf{R}^n$, $K \subset U$ a compact set, and $V \supset K$ an open set strictly interior to $U$. Then there is a constant $\delta \in (0, 2)$ such that

(4.1) $$|f(x_1) - f(x_2)| \le C_n M T \theta(|x_1 - x_2|)$$

for every mapping $f : U \to \mathbf{R}^m$ which belongs to the class $W^1_{n,loc}(U)$ and satisfies the condition $E(T)$ and for any two points $x_1, x_2 \in K$ with $|x_1 - x_2| < \delta$. Here $C_n$ is a constant which depends only on $n$, and

$$M = \|f\|_{L^1_n(V)}, \quad \theta(t) = (2/\ln(2/t))^{1/n}, \quad 0 < t < 2, \quad \theta(0) = 0.$$

For the proof of Lemma 2 see [5].

Here we shall consider the functionals of the variational calculus of the form

$$(4.2) \qquad I_{F,U}(f) = \int_U F(x, f'(x))dx,$$

where $U$ is an open set in $\mathbf{R}^n$, $f : U \to \mathbf{R}$ is a real function and $f'(x)$ denotes the gradient of the function. It will be supposed that the function $F$ here satisfies the conditions F1, F2 and F3 given below.

F1. There exists a set $E \subset U$ with meas $E = 0$ such that, for every $x \in U \setminus E$, $F(x, p)$, is defined as a function of the variable $p$ in all of $\mathbf{R}^n$, has partial derivatives $\frac{\partial F}{\partial p_i}(x, p)$ for every $p \in \mathbf{R}^n$, and is convex as a function of $p$.

F2. For every bounded open $V \subset U$ and for every $\varepsilon > 0$ one can find a compact $A \subset V$ such that meas $(V \setminus A) < \varepsilon$ and the functions $F$, $\frac{\partial F}{\partial p_i}$, $i = 1, 2, \ldots, n$ are continuous on $A \times \mathbf{R}^n$.

Let

$$F_p(x, p) = \left( \frac{\partial F}{\partial p_1}(x, p), \frac{\partial F}{\partial p_2}(x, p), \ldots, \frac{\partial F}{\partial p_n}(x, p) \right).$$

F3. For every $x \in U \setminus E$, where $E$ is the set introduced in F1, for every $p \in \mathbf{R}^n$

$$(4.3) \qquad \lambda_0 |p|^n \le F(x, p),$$

$$(4.4) \qquad \lambda_1 |p|^n \le \langle F_p(x, p), p \rangle,$$

$$(4.5) \qquad \lambda_2 |p|^{n-1} \ge |F_p(x, p)|,$$

and the function $c(x) \equiv F(x, 0)$ is locally integrable in $U$.

From (4.5) one can deduce that

$$(4.6) \qquad F(x, p) \le c(x) + \lambda_2 |p|^n$$

for every $x \in U \setminus E$.

Condition F2 allows us to conclude that for every measurable vector function $p(x)$ in $U$ the functions $x \mapsto F(x, p(x))$, $x \mapsto \frac{\partial F}{\partial p_i}(x, p(x))$ will be measurable in $U$. The estimates for the condition F3 lead to the conclusion that for every function $f : U \to \mathbf{R}$ of the class $W^1_{n,loc}(U)$ the function $x \mapsto F(x, f'(x))$ will be locally summable in $U$ and the functions $\frac{\partial F}{\partial p_i}(x, f'(x))$ will be locally summable of degree $\frac{n}{n-1}$.

Let $I_{F,U}$ be a functional given by the equality (4.2). We shall say that $I_{F,U}$ is normal if the function $F$ satisfies conditions F1, F2, F3.

Let $f$ be an arbitrary function of the class $W^1_{n,loc}(U)$. We shall say that $f$ is a stationary function for the normal functional $I_{F,U}$ if for every function $\varphi : \mathbf{R}^n \to \mathbf{R}$ of the class $C^\infty(\mathbf{R}^n)$ with support contained in $U$

$$(4.7) \qquad \int_U \langle F_p(x, f'(x)), \varphi'(x) \rangle dx = 0.$$

By simple arguments based on passage to the limit one can show that if $f(x)$ is a stationary function for the normal functional $I_{F,U}$ then equality (4.7) will be true for every function $\varphi$ of the class $W_n^1(U)$ with compact support . (It is supposed, of course, that $F$ satisfies conditions F1, F2, F3.)

Let us fix the constants $\lambda_0, \lambda_1, \lambda_2$ and let $\mathfrak{G}(\lambda_0, \lambda_1, \lambda_2)$ be the set of all functions $f : U \to \mathbf{R}$ with the following property. If the function $f$ belongs to $\mathfrak{G}(\lambda_0, \lambda_1, \lambda_2)$ then it is a stationary function for a functional $I_{F,U}$ of the form (4.2) with $F$ satisfying all conditions F1, F2, F3. The constants in the formulation of the condition F3 must be the same as in the expression $\mathfrak{G}(\lambda_0, \lambda_1, \lambda_2)$.

If the function $f(x)$ of the class $W_{n,loc}^1(U)$ is the stationary function for a functional of type (4.2) then $f(x)$ satisfies a Hölder condition on every compact subset of $U$. This fact is a particular case of an essentially more general theorem. In the general case all known proofs are very complicated (see, for instance, [3], [4], [5]). For the special case considered here a simplified proof of the Hölder continuity of stationary functions was given also in [7]. In [7] the Hölder continuity was deduced from the Harnack inequality. Here we show how to reach the same goal using Theorem 1.

Obviously the class $\mathfrak{G}(\lambda_0, \lambda_1, \lambda_2)$ satisfies condition K2.

Let $f$ be the stationary function of the functional $I_{F,U}$ with $F$ satisfying conditions F1–F3. Let $g = \lambda f + a$ where $\lambda > 0, \lambda, a$ are constants. Let $F^*(x,p) = \lambda^n F(x, \frac{p}{\lambda})$. Then $F^*(x,p) = \lambda^n F(x, \frac{p}{\lambda}) \geq \lambda^n \lambda_0 \frac{|p|^n}{\lambda^n} = \lambda_0 |p|^n$. Further,

$$F_p^*(x,p) = \lambda^{n-1} F_p(x, \frac{p}{\lambda}),$$

from which it follows immediately that

$$\langle F_p^*(x,p), p \rangle = \lambda^n \langle F_p(x, \frac{p}{\lambda}), \frac{p}{\lambda} \rangle \geq \lambda_1 |p|^n,$$

$$|F_p^*(x,p)| \leq \lambda_2 |p|^{n-1}.$$

So we see that all inequalities (4.3), (4.4), and (4.5) are all true for $F^*(x,p)$. For every $\varphi \in C^\infty(\mathbf{R}^n)$ with compact support in $U$ we have

$$\int_U \langle F_p^*(x, g'(x)), \varphi'(x) \rangle dx = \lambda^n \int_U \langle F_p(x, f'(x)), \varphi'(x) \rangle dx = 0.$$

This means that $g$ is the stationary function for the functional $I_{F^*,U}$, and hence the class $\mathfrak{G}(\lambda_0, \lambda_1, \lambda_2)$ is invariant with respect to transformations of the form $f \mapsto \lambda f + a$.

Let $\lambda > 0$ be a real number, let $a$ be a vector in $\mathbf{R}^n$, and let $g(x) = f(\lambda x + a)$. Let $U^*$ be the image of $U$ under the homothety $x \mapsto (x - a)/\lambda$. Then for $x \in U^*$, we have $\lambda x + a \in U$. Let

$$F^*(x,p) = \lambda^n F(\lambda x + a, \frac{p}{\lambda}).$$

Then by the same arguments as above it is established that $F^*$ satisfies conditions F1–F3 and that $g$ is the stationary function for the functional $I_{F^*,U^*}$. Then since the function $f \in \mathfrak{G}(\lambda_0, \lambda_1, \lambda_2)$ is arbitrary it follows that the class $\mathfrak{G}(\lambda_0, \lambda_1, \lambda_2)$ is also invariant relative to the transformation $f \mapsto f \circ \alpha$ where $\alpha$ is any homothety in $\mathbf{R}^n$.

Every stationary function of a normal functional, as it is shown for instance in [5], satisfies the maximum and minimum principles. This means that for every such function the condition $E(T)$ with $T = 1$ of Lemma 2 is true. From this lemma it thus follows that all stationary functions of normal functionals are continuous functions.

Let $\mathfrak{G}_0$ be the set of all functions belonging to the class $\mathfrak{G}(\lambda_0, \lambda_1, \lambda_2)$ which are defined on the unit ball $B(0,1)$ and satisfy the following conditions: $f(0) = 0$ and $|f(x)| \leq 1$ for all $x \in B(0,1)$. Our goal is to show that $\mathfrak{G}(\lambda_0, \lambda_1, \lambda_2)$ is a class of type $K(n,1)$. This will be achieved if we show that for the set of functions $\mathfrak{G}_0$ the condition K3 is satisfied. We show that the class $\mathfrak{G}_0$ is bounded in the space $W_n^1(B(0,2/3))$. In view of Lemma 2 this result allows us to conclude that the class $\mathfrak{G}_0$ is uniformly equicontinuous on the ball $\overline{B}(0,1/2)$ (in fact on every ball $\overline{B}(0,\rho)$, where $0 < \rho < \frac{2}{3}$).

Let $f$ be an arbitrary function of the class $\mathfrak{G}_0$. This means that $f$ is the stationary function for a normal functional $I_{F,U}$ with $U = B(0,1)$. Let $\zeta$ be a real function of the class $C^\infty(\mathbf{R}^n)$ such that $\zeta(x) = 1$ for $|x| \leq \frac{2}{3}$, $\zeta(x) = 0$ for $|x| > \frac{3}{4}$, and $0 \leq \zeta(x) \leq 1$ for all $x \in \mathbf{R}^n$. Setting $U = B(0,1)$ and $\varphi = \zeta^n f$ in the equality (4.7) we get

$$0 = \int_{B(0,1)} \langle F_p(x, f'(x)), f'(x) \rangle \zeta^n(x) dx$$

$$+ n \int_{B(0,1)} \langle F_p(x, f'(x)), \zeta'(x) \rangle \zeta^{n-1}(x) f(x) dx.$$

Consequently

$$(4.8) \int_{B(0,1)} \langle F_p(x, f'(x)), f'(x) \rangle \zeta^n(x) dx = -n \int_{B(0,1)} \langle F_p(x, f'(x)), \zeta'(x) \rangle \zeta^{n-1}(x) f(x) dx$$

$$\leq n \int_{B(0,1)} |F_p(x, f'(x))| |\zeta'(x)| \zeta^{n-1}(x) dx.$$

Here we have used the fact $|f(x)| \leq 1$ for every $x \in B(0,1)$. Using the inequalities (4.4) and (4.5) we get from (4.8):

$$\lambda_1 \int_{B(0,1)} |f'(x)|^n \zeta^n(x) dx \leq n\lambda_2 \int_{B(0,1)} |f'(x)|^{n-1} (\zeta(x))^{n-1} |\zeta'(x)| dx.$$

In the Young inequality

$$uv \leq \frac{(n-1)u^{\frac{n}{n-1}}}{n\xi} + \frac{\xi^{n-1}v^n}{n}$$

if we set $u = |f'(x)|^{n-1}(\zeta(x))^{n-1}$, $v = |\zeta'(x)|$ we arrive at the inequality

$$(4.9) \qquad \int\limits_{B(0,1)} |f'(x)|^n \zeta^n(x) dx \leq \frac{(n-1)\lambda_2}{\lambda_1 \xi} \int\limits_{B(0,1)} |f'(x)|^n [\zeta(x)]^n dx$$

$$+ \frac{\xi^{n-1}}{\lambda_1} \int\limits_{B(0,1)} |\zeta'(x)|^n dx.$$

Setting $\xi = 2(n-1)\lambda_2/\lambda_1$, after some trivial transformations we get

$$\int\limits_{B(0,2/3)} |f'(x)|^n dx \leq \int\limits_{B(0,1)} |f'(x)|^n [\zeta(x)]^n dx \leq C = \frac{2\xi^n}{\lambda_1} \int\limits_{B(0,1)} |\zeta'(x)|^n dx = \text{const.}$$

The boundedness of the class $\mathfrak{G}_0$ in the space $W_n^1(B(0,2/3))$ is proved. Thus condition K3 is satisfied for the class $\mathfrak{G}_0$ and we see that the class $\mathfrak{G}(\lambda_0, \lambda_1, \lambda_2)$ is a class of type $K(n,1)$.

Applying Theorem 1 we see that every stationary function of a normal functional $I_{F,U}$ satisfies a Hölder condition on any compact subset of the domain where this stationary function is defined. This is what was required to be proved.

## 5. Concluding remarks

In conclusion we shall make some short remarks about the application of Theorem 1 to quasiconformal and quasiregular mappings. By using this theorem we can establish the Hölder continuity of such mappings at the beginning stage of their study. For this goal it is necessary to establish at first the following properties of such mappings.

I) Every quasiconformal and every quasiregular mapping satisfies the condition $E(T)$ for some $T$ (in fact, one can take $T = 1$).

II) Every family of quasiregular mappings $f$ such that $|f(x)| \leq 1$ for all $x \in B(0,1)$ is bounded in the space $W_n^1(B(0,d))$ for some $d \in (0,1]$.

The proofs of these two assertions do not require the use of any deep results of the theory of mappings with bounded distortion (see, for instance [5]).

Finally we remark that, in view of Theorem 1, Hölder continuity of quasiconformal mappings follows from an old result of M.A. Kreines.

**Added in proof.** M. Vuorinen has called my attention to the fact that a proof of the Hölder continuity completely analogous to the proof of Theorem 1 but for a different class of functions was given by Tukia and Väisälä [8].

# References

[1]     A. P. Kopylov, Stability of classes of multidimensional holomorphic mappings I. The concept of stability. Liouville's theorem, Sibirsk. Mat. Zh. Vol. 23, No. 3 (1982), 70–91.

[2]     A. P. Kopylov, Stability in the $C$-norm of classes of mappings, Novosibirsk "Nauka", Siberian subdivision, 1990, 1–222.

[3]     J. Moser, On Harnack's theorem for elliptic differential equations, Comm. Pure Appl. Math. 14 (1961), 577–591.

[4]     O. A. Ladyzhenskaya and N. N. Ural'tseva, Linear and quasilinear elliptic equations, Moscow "Nauka", 1973.

[5]     Yu. G. Reshetnyak, Space mappings with bounded distortion, Translations of mathematical monographs v 73, American math. society, 1989, revised and enlarged translation of "Prostranstvennye otobrazheniya s ogranichennym iskazheniem", Novosibirsk "Nauka", 1982.

[6]     M. A. Kreĭnes, Sur une classe de fonctions de plusier variables, Mat. Sb. 9 (51) (1941), 713–720.

[7]     S. Granlund, P. Lindqvist, and O. Martio, Conformally invariant variational integrals, Trans. Amer. Math. Soc. 277 (1983), 43–73.

[8]     P. Tukia and J. Väisälä: Quasisymmetric embeddings of metric spaces, Ann. Acad. Sci. Fenn. Ser. A I Math. 5 (1980), 97–114.

Quasiconformal Space Mappings
– A collection of surveys 1960–1990
Springer–Verlag (1992), 93–103
Lecture Notes in Mathematics Vol. 1508

# PICARD'S THEOREM AND DEFECT RELATION FOR
# QUASIREGULAR MAPPINGS

Seppo Rickman
University of Helsinki, 00100 Helsinki, Finland

## 1. Introduction

Quasiregular mappings were introduced in 1966 by Yu. G. Reshetnyak. The definition is obtained from that of quasiconformal mappings by leaving out the homeomorphism requirement. More precisely, we say that a continuous mapping $f : G \to R^n$ of a domain $G$ in $R^n$ is quasiregular $(qr)$ if

$$(1.1) \qquad f \in W^1_{n,loc}(G),$$

$$(1.2) \qquad \text{there exists } K, \ 1 \leq K < \infty, \text{ such that } |f'(x)|^n \leq KJ_f(x) \text{ a.e.}.$$

Here $f \in W^1_{n,loc}(G)$ means that $f$ belongs locally to the Sobolev space of maps in $L^n$ with weak first order partial derivatives which are in $L^n$. Furthermore, $f'(x)$ is the formal derivative defined in terms of the partial derivatives, $J_f(x)$ is the Jacobian determinant of $f$ at $x$, and $|f'(x)|$ is the operator norm of $f'(x)$. The smallest $K$ is (1.2) is the outer dilatation $K_O(f)$, and the smallest $K' \in [1, \infty[$ in

$$(1.3) \qquad\qquad J_f(x) \leq K' \min_{|h|=1} |f'(x)h|^n$$

is the inner dilatation $K_I(f)$. The (maximal) dilatation of $f$ is $K(f) = \max(K_O(f), K_I(f))$ and we call a $qr$ map $f$ $K$–quasiregular if $K(f) \leq K$. The definition extends as such to oriented Riemannian $n$–manifolds. Sometimes we call a $qr$ map $f : G \to S^n$ quasimeromorphic $(qm)$. A (sensepreserving) quasiconformal $(qc)$ map is a $qr$ homeomorphism. Part of the basic theory was given in the articles [MRV 1–3] by O. Martio, S. Rickman, and J. Väisälä.

The interest in $qr$ maps is due to the fact that they give a natural generalization of the geometric part of the theory of (complex) analytic functions to real $n$–dimensional space. The purpose of this survey is to describe some geometric aspects and to give a summary of results about value distribution of $qr$ maps. It turns out that analogues exist for various classical results starting from Picard's theorem on omitted values and including a defect relation in the spirit of Ahlfors's theory.

From another point of view the study of $qr$ maps is closely connected to *PDE* theory. For example, as observed by Reshetnyak, the coordinate functions $f_j$ of a $qr$ map satisfy a second order *PDE* of divergence form, which is the Euler–Lagrange equation of a variational integral which is invariant under conformal change of metric. Solutions of such equations give $n$–dimensional counterparts for harmonic functions in dimension two. One of Reshetnyak's main result, that a nonconstant $qr$ map is discrete and open, is proved by means of such a potential theory. Martio and his students have extended the study of this nonlinear potential theory in the Euclidean case. Their treatment also includes counterparts of subharmonic functions. See the article [M] by Martio in this volume. On Riemannian $n$–manifolds such a potential theory is studied in [H1], [HR1] and [H2]. It was shown recently by T. Iwaniec and G. Martin [IM] in even dimensions, and by Iwaniec [I] in all dimensions, that the ideas can be extended via Hodge theory to produce efficient equations for $\ell$–forms, too, $(1 \leq \ell < n)$ in connection with $qr$ maps. They were able to solve affirmatively a longstanding problem on removability of sets with positive Hausdorff dimension.

Two monographs on $qr$ maps have appeared, namely, one by Reshetnyak [Re] and another by M. Vuorinen [Vu1]. A third one [R12] is in preparation.

## 2. Picard type theorems

At an early stage of the development of the theory of $qr$ maps V.A. Zorich published an important result in [Z] which says that a locally $qr$ map of $R^n$ into itself is in fact a homeomorphism, hence $qc$, provided $n \geq 3$. In the same paper he gave an example of a nonconstant $qr$ map $f : R^n \to R^n \setminus \{0\}$, which can be regarded as a counterpart of the planar exponential function $z \mapsto e^z$ in complex notation. He also posed the question of the validity of a counterpart of Picard's theorem that a (complex) analytic function of the plane into itself omitting at least two points must be constant. Note that in the plane a $qr$ map $f$ is of the form $f = \varphi \circ h$, where $h$ is $qc$ and $\varphi$ analytic. Hence Picard's theorem follows for $qr$ maps in dimension two from this representation.

For a period of time it was conjectured that a Picard theorem would be true for $qr$ maps for $n \geq 3$ in the same form as in the plane. This turned out to be false, at least in dimension three. For all dimensions $n \geq 3$ the following Picard type theorem was proved in [R4]:

**2.1. Theorem.** *For each $K > 1$ there exists an integer $q = q(n, K)$ such that every $K - qr$ map $f : R^n \to R^n \setminus \{a_1, \ldots, a_q\}$, where $a_1, \ldots, a_q \in R^n$ are distinct, is constant.*

That this is qualitatively best possible in dimension three is shown by the following result.

**2.2. Theorem.** [R8] *For each positive integer $p$ there exists a nonconstant $K(p)-qr$ map $f : R^3 \to R^3$ omitting $p$ points.*

By now there exist several different proofs for Theorem 2.1. The original proof combines the method of extremal length with the potential theory described in the introduction. The essential tool is a sharp comparison lemma on averages of covering numbers over $(n-1)$-spheres which is proved by means of an inequality on moduli of path families. The inequality is due to Väisälä [V2] and is based on ideas by E.A. Poletskiĭ [P]. Proofs depending only on extremal length methods have been given in [R6], [R7], [R10], [R12]. The proofs give immediately a counterpart of the big Picard theorem too concerning the removability of an isolated singularity. Recently, A. Eremenko and J.L. Lewis [EL] gave a purely potential theoretic proof for Theorem 2.1. We shall in 2.6 sketch the idea of the proof by means of extremal length.

Compared to the various proofs of 2.1 the proof of Theorem 2.2 is extremely technical. One can in fact show that as soon as $p \geq 2$ the map $f$ must be of complicated nature. It is likely that the result in 2.2 is true also for $n \geq 4$ but the details have not been worked out.

A quantitative result which corresponds to the classical Picard–Schottky theorem is obtained too. It can be formulated as follows.

**2.3. Theorem.** [R7] *There exists a constant $\delta = \delta(n, K) > 0$ such that each $K - qr$ map $f : B^n \to S^n \setminus \{a_1, \ldots, a_{q+1}\}$ satisfies*

$$(2.4) \qquad \tau(f(x), f(y)) \leq C \max\,(\rho(x,y), \delta).$$

*Here $q = q(n, K)$ is the integer from Theorem 2.1, $\rho$ is the Poincaré distance in the unit ball $B^n$, and $\tau$ is the distance for a metric $d\tau$ in $S^n \setminus \{a_1, \ldots, a_{q+1}\}$, which is conformal with respect to the standard metric $d\sigma$ in $S^n$ with ratio*

$$(2.5) \qquad \frac{d\tau}{d\sigma} = \frac{1}{\sigma(a_j, x)|\log \sigma(a_j, x)|}$$

*near $a_j$. The constant $C$ depends only on $n, K$, and the minimal distance between the points $a_j$.*

The reader will recognize (2.5) to be the order of ratio for the hyperbolic metric in a 2-sphere with punctures. Theorem 2.3 is therefore the true analogue of the distance decreasing result for analytic functions of the disk into $S^2 \setminus \{a_1, \ldots, a_{q+1}\}$, $q \geq 2$, when the latter is provided with its hyperbolic metric.

**2.6. Extremal length proof of Theorem 2.1.** We shall give an outline of the proof which involves only inequalities of moduli of path families. The main idea is to show that too many omitted points result in a rapid growth relation on a measure which is defined by means of the average covering number over $(n-1)$-spheres. To be more specific, let now $f : R^n \to R^n \setminus \{a_1, \ldots, a_q\} = Y$ be a nonconstant $K - qr$ map and let us assume for notational convenience that $a_1, \ldots, a_q \in B^n(1/2)$. For any point $y \in Y$ and any Borel set $E \subset R^n$ let $n(E, y)$ be the number of points in $f^{-1}(y) \cap E$ with multiplicity (which has a definite meaning for discrete open maps) regarded. If $E = \overline{B}^n(r)$, we also write $n(r, y)$ for $n(E, y)$. Then let $\nu(E, X)$ be the average of $n(E, y)$ over an $(n-1)$-sphere $X$ with respect to the $(n-1)$-measure on $X$ induced from $R^n$. The comparison lemma mentioned above has the following form.

**2.7. Lemma.** *For $r, s, t > 0$ and $\theta > 1$ we have*

$$(2.8) \qquad \nu(\theta r, t) \geq \nu(r, s) - \frac{K_I |\log \frac{t}{s}|^{n-1}}{|\log \theta|^{n-1}},$$

*where we have written $\nu(\rho, a) = \nu(\overline{B}^n(\rho), S^{n-1}(a))$.*

From this lemma we see that the average $\nu(\theta r, t)$ is comparable well with $\nu(r, s)$ from below if max $(t/s, s/t)$ is small or $\theta$ is large. The reader may check the sharpness of (2.8) with the map $z \mapsto z^k$ in the plane. A general theory on averages on Riemannian manifolds in connection with $qr$ maps is given in [MR].

To start the proof of 2.1 we make the preliminary observation that any average $\nu(r, s)$ tends to $\infty$ as $r \to \infty$ if $f$ is nonconstant and omits at least one point in $R^n$. This is a simple consequence of a result that such a map must fill the space except possibly a set of $n$-capacity zero in each neighborhood of an essential singularity. A compact set $E$ in $R^n$ is by definition of zero $n$-capacity if for all open $U \subset R^n$ such that $E \subset U$ we have

$$\inf_u \int_U |\nabla u|^n = 0$$

where $u$ runs over the set of all functions in $C_0^\infty(U)$ with $u \mid E = 1$. A closed set is of zero $n$-capacity if each of its compact subsets is.

Now let $\nu(r, 1)$ be large in a sense which will be clear along the proof. We apply 2.7 together with an auxiliary $qc$ map to get

$$(2.9) \qquad \nu(2r, S^{n-1}(a_j, \sigma)) \geq \frac{1}{2}\nu(r, 1),$$

where $\sigma > 0$ is given by an equation

$$(2.10) \qquad (\log \frac{1}{\sigma})^{n-1} = c_1 \nu(r, 1).$$

Here and in the following $c_1, c_2, \ldots$ are positive constants which depend only on $n$ and $K$. The inequality means in particular that $f^{-1}\overline{B}^n(a_j, \sigma)$ meets $\overline{B}^n(2r)$ for all $j$. Since $a_j$ is omitted and $f$ is discrete open, each component of $F_j = f^{-1}\overline{B}^n(a_j, \sigma)$ tends to infinity. Let $\Gamma_j$ be the family of paths in $B^n(3r) \setminus \overline{B}^n(2r)$ that connect $F_j$ to $\cup_{k \neq j} F_k$. Then each image path $f \circ \gamma$, $\gamma \in \Gamma_j$, connects the spheres $S^{n-1}(a_j, \sigma)$ and $S^{n-1}(a_j, \sigma_0)$ where $\sigma_0 = \frac{1}{4} \min |a_k - a_m|$. Note that $\sigma$ is small enough so that $\sigma < \sigma_0$. The function $\rho$ defined by

$$\rho(y) = \frac{1}{(\log \frac{\sigma_0}{\sigma})|y|}$$

in the ring $A = B^n(a_j, \sigma_0) \setminus \overline{B}^n(a_j, \sigma)$ and zero elsewhere is admissible for the image family $f\Gamma_j$, i.e. it is a nonnegative Borel function and the line integral satisfies

$$\int_{f \circ \gamma} \rho ds \geq 1$$

for $\gamma \in \Gamma_j$. The $n$-modulus of $f\Gamma_j$ is obtained as

$$M_n(f\Gamma_j) = \inf_{\eta} \int \eta^n dm$$

where the infimum is taken over all admissible $\eta$ for $f\Gamma_j$. For properties of the $n$-modulus, see [V1]. A transformation formula for integrals then gives via the spherical Fubini theorem the estimate

(2.11)
$$M_n(\Gamma_j) \leq K \int_A \rho(y)^n n(3r, y) dy$$

$$\leq \frac{c_2}{(\log \frac{\sigma_0}{\sigma})^n} \int_\sigma^{\sigma_0} \frac{\nu(3r, S^{n-1}(a_j, t))}{t} dt.$$

Next we use the comparison lemma 2.7 again and estimate

$$\nu(3r, S^{n-1}(a_j, t)) \leq \nu(4r, 1) + c_3 (\log \frac{1}{\sigma})^{n-1}.$$

Putting this into (2.11) yields

(2.12)
$$(M_n(\Gamma_j) - c_4)(\log \frac{1}{\sigma})^{n-1} \leq c_5 \nu(4r, 1).$$

By standard estimates on the $n$-modulus we can deduce that $\max_j M_n(\Gamma_j) \to \infty$ as $q \to \infty$. From (2.10) and (2.12) we then conclude that there exists $q = q(n, K)$ such that

$$\frac{\nu(4r, 1)}{\nu(r, 1)} \geq \alpha_n$$

where $\alpha_n$ is so large that there exists a ball $\overline{B}^n(x_1, r/2)$, $x_1 \in B^n(4r)$, where the measure $E \mapsto \nu(E, 1)$ exceeds $\nu(r, 1)$, i.e. $\nu(\overline{B}^n(x_1, r/2), 1) \geq \nu(r, 1)$. Then we can repeat by starting from $\overline{B}^n(x_1, r/2)$ instead of $\overline{B}^n(r)$. This procedure converges to a point $x_0$ where we would have $\nu(\{x_0\}, 1) > 0$, which is a contradiction. □

As can be seen from above the main idea of the proof is not very complicated. However, the comparison lemma 2.7 involves rather technical details in the proof of the so called Poletskiǐ's lemma [P] which is contained in the proof of Väisälä's inequality [V2]

(2.13)
$$M_n(\Gamma') \leq \frac{K_I}{m} M_n(\Gamma).$$

In (2.13) $\Gamma'$ is a family in the target of $f$ and the assumption is roughly that every path $\beta \in \Gamma'$ has $m$ partial lifts $\alpha_1, \ldots, \alpha_m$ in $\Gamma$.

For these technical reasons the contribution by Eremenko and Lewis in [EL], where they present a purely potential theoretic proof, is noteworthy. Some time ago M. Gromov posed the question whether Theorem 2.1 remains true if $Y = R^n \setminus \{a_1, \ldots, a_q\}$ is given an arbitrary Riemannian metric. I. Holopainen gave a partial affirmative solution to this question in his thesis [H1]. He puts a condition on the metric which ensures that certain path families can be constructed. It turns out that using ideas from Holopainen's paper [H1] to construct certain potentials in $Y$ and combining this with ideas by Eremenko and Lewis we can give an affirmative answer to Gromov's question. Details are presented in [HR2]. Generalizations in another direction are given by M. Vuorinen in [Vu2] and by P. Järvi in [J], where they study sets called Picard sets in classical theory, for qr maps.

## 3. Remarks on existence of quasiregular mappings

Theorem 2.1 is an example of the general question of the existence of a nonconstant qr map $f : M \to N$ when two Riemannian $n$–manifolds $M$ and $N$ are given. Not very much is known about this in general. In particular, one would like to find good geometric invariants for this question. Let us consider some cases where the answer is known.

First let $M = R^n$ and let $N$ be a compact $n$–manifold. For $n = 3$ we can study what the answer is if $N$ is provided by one of the eight geometries listed by W. Thurston. It turns out (see [Jo]) that only in the three cases where the universal cover $\tilde{N}$ is $R^3$, $S^3$ or $R^1 \times S^2$, there exists a nonconstant qr map of $R^3$ into $N$. In all other five cases the 3–capacity at infinity of $\tilde{N}$ is positive, which fact easily gives the nonexistence.

Of particular interest is the geometry of the Heisenberg group $H_1$, which is the nilpotent Lie group, presented as a matrix group by

$$H_1 = \left\{ \begin{pmatrix} 1 & x & z \\ 0 & 1 & y \\ 0 & 0 & 1 \end{pmatrix} : x, y, z \in R^1 \right\}.$$

Equipped with a left invariant Riemannian metric $H_1$ satisfies an isoperimetric inequality of the form

$$(3.1) \qquad \text{vol}\,(A) \le C \mathcal{H}^{n-1}(\partial A)^{\frac{m}{m-1}}$$

with $m = 4$, $n = 3$, and $A \subset\subset H_1$. $\mathcal{H}^{n-1}(\partial A)$ is the $(n-1)$–dimensional Hausdorff measure of $\partial A$. The proof of (3.1) is given in [Pa1]. In general, as soon as one has (3.1) with $m$ exceeding the dimension, then the manifold has positive $n$–capacity at infinity [Pa1], see too [G, p. 86]. Holopainen proved in [H1] directly that $H_1$ has positive 3–capacity at infinity. Looking at qr maps of $H_1$ we have the following information. There exists a nonconstant qr map of $H_1$ into $R^3$ [HR3]. There exists no such map of $H_1$ into the hyperbolic 3–space [H2] (this result was announced orally to me by M. Gromov), or even more, there exists a Picard type theorem for maps of $H_1$ [HR3].

Let us continue with the case $M = R^n$, $N$ compact. K. Peltonen [Pel] has proved for example the following result: If $N$ is a connected sum $T^n \# P$ of the $n$–torus and a compact Riemannian $n$–manifold $P$ with a nontrivial cohomology group $H^m(P)$ for some $m \in ]0, n[$, then there is no nonconstant qr map of $R^n$ into $N$. After some preliminary steps, where both the property $\pi_1(T^n) = Z^n$ and the nontriviality of $H^m(P)$ are used, the proof is a modification of the one presented in 2.6. A typical example for $N$ is $T^4 \# S^2 \times S^2$. It is easy to construct a nonconstant qr map of $R^4$ onto $S^2 \times S^2$, namely, first by projection onto $T^4 = T^2 \times T^2$ and then by mapping each factor $T^2$ onto $S^2$

by a 2 to 1 branched covering map. An interesting open question is whether the same statement is true for $S^2 \times S^2 \# S^2 \times S^2$.

If $M$ is any oriented Riemannian $n$–manifold, there always is a nonconstant $qr$ map of $M$ onto $S^n$. This is proved in [Pel] too. The main task is to construct a triangulation in $M$ whose $n$–simplices can be mapped by a $L$–bilipschitz map followed by a homothety onto a standard $n$–simplex with fixed $L$. Sometimes such a triangulation is called fat in the literature.

## 4. Defect relation

For meromorphic functions in the plane R. Nevanlinna established his famous value distribution theory in 1925 [N1]. It is a far reaching generalization of Picard's theorem in the sense that it gives precise information of how much less given fixed points can be covered compared to the average covering number. This information is contained in the so called Second Main Theorem in his theory. It implies for example Nevanlinna's defect relation which says that the sum of defects $\delta_a$, $a \in \overline{R}^2$, is bounded by 2. The defect $\delta_a$ measures how much less in general the point $a$ is covered compared to the average covering number. For details, see [N1] or [N2]. In 1935 L.V. Ahlfors established a parallel theory in his celebrated paper [A] on covering surfaces. His theory is more geometric than the one by Nevanlinna and it is better suited for generalizations. It was J. Miles who showed in [Mi1] that most of Nevanlinna's results can be deduced from those of Ahlfors. One form of a defect relation in Ahlfors's theory can be formulated as follows. Let $f : R^2 \to \overline{R}^2$ be a nonconstant meromorphic function. Then there exists a set $E \subset [1, \infty[$ of finite logarithmic measure, i.e.

$$(4.1) \qquad \int_E \frac{dr}{r} < \infty,$$

such that

$$(4.2) \qquad \limsup_{E \not\ni r \to \infty} \sum_{j=1}^{q} (1 - \frac{n(r, a_j)}{A(r)})_+ \leq 2$$

whenever $a_1, \ldots, a_q$ are distinct points in $\overline{R}^2$. Recall the notation $n(r, y)$ from 2.6. Here $A(r)$ is the average of $n(r, y)$ with respect to the spherical 2–measure of $\overline{R}^2$ and $\alpha_+$ is $\max(\alpha, 0)$ for $\alpha \in R^1$.

If we write

$$(4.3) \qquad \delta_f(a_j, r) = \delta(a_j, r) = (1 - \frac{n(r, a_j)}{A(r)})_+$$

and call this the *defect function* of $a_j$, we see from (4.2) that the total contribution of any number of defect functions is bounded by 2 provided $r$ stays out a thin exceptional set $E$. Picard's theorem on omitted values is a trivial consequence of (4.2).

We may now ask if there is a counterpart of (4.2) for $qr$ maps when $n \geq 3$. For $n = 2$ this was settled already by Ahlfors in [A]: the same inequality (4.2) is true for such maps. Let now $f : R^n \to \overline{R}^n$ be $K - qr$ and nonconstant. In [R] a defect relation was proved in the form

$$(4.4) \qquad \limsup_{E \not\ni r \to \infty} q\left(\frac{1}{q} \sum_{j=1}^{q} \delta(a_j, r)^{n-1}\right)^{\frac{1}{n-1}} \leq C(n, K)$$

where $C(n, K)$ is a finite constant depending only on $n$ and $K$ and the exceptional set $E$ satisfies again (4.1). Although the left hand side of (4.4) reduces to that of (4.2) for $n = 2$, the inequality is not best possible. Finally, in [R11] (see also [R12] for a complete proof) the right form for the defect relation was obtained and it turns out that the left hand side of (4.4) can indeed be replaced by that in (4.2):

**4.5. Theorem.** *Let* $f : R^n \to \overline{R}^n$ *be* $K - qr$ *and nonconstant. Then there exists a set* $E \subset [1, \infty[$ *satisfying (4.1) such that*

$$(4.6) \qquad \limsup_{E \not\ni r \to \infty} \sum_{j=1}^{q} \delta(a_j, r) \leq C(n, K)$$

*whenever* $a_1, \ldots, a_q$ *are distinct points in* $\overline{R}^n$.

In the classical theory of meromorphic functions it was a long standing open problem whether arbitrarily given Nevanlinna defect numbers $\delta_{a_j}$, $j = 1, 2, \ldots$, attached to given points $a_j$ in $\overline{R}^2$ and subject only to the condition

$$\sum_j \delta_{a_j} \leq 2$$

given by the theory, can be realized by a meromorphic function. The full solution to this problem was given by D. Drasin in [D]. In fact, his function isa realization for the so called ramification indices too, appearing in the Nevanlinna theory.

The same question arises naturally for $qr$ maps when $n \geq 3$. It is now known for $n = 3$ [R11] that arbitrary given asymptotic values for the defect functions subject only to a bound $P$ can be realized by a $K(P) - qr$ map $f : R^3 \to \overline{R}^3$ outside an exceptional set. More precisely, we have the following result:

**4.7. Theorem.** *Let* $a_1, a_2, \ldots$ *be a sequence of points in* $\overline{R}^3$ *and let* $0 \leq \delta_1, \delta_2, \ldots \leq 1$ *be numbers such that*

$$\sum_j \delta_j \leq P.$$

*Then there exists a* $K(P) - qr$ *map* $f : R^3 \to \overline{R}^3$ *and a set* $E \subset [1, \infty[$ *satisfying (4.1) such that*

$$\lim_{E \not\ni r \to \infty} \delta_f(a_j, r) = \delta_j,$$

$$\lim_{E \not\ni r \to \infty} \delta_f(y, r) = 0, \quad y \notin \{a_1, a_2, \ldots\}.$$

Theorem 4.7 shows that in a sense the defect relation 4.5 is best possible. With some extra effort it is possible to establish the limits in Theorem 4.7 wihtout the exceptional set $E$ (Drasin's result in [D] for meromorphic functions is of this type). For $P = 2$ such a result was shown in [R1] for $n = 3$ and by the help of techniques from [R3] for $n \geq 3$. The proof of 4.5 is put together via a careful study of lifts of paths and it is in the spirit of the proof of Theorem 2.1 presented here in 2.6, but much more involved. A modification of the proof in the plane gives a new proof of Ahlfors's result (4.2) as well. This was shown by M. Pesonen [Pes], see too [R12, Chapter V]. The proof of 4.7 is based on ideas from the proof of 2.2 and [R1], but again the technical details are much worse.

**4.8. Concluding comments.** Although Picard type theorems and the defect relation are known in all dimensions and their sharpness in dimension three, there is much left for further research.

We already discussed the general existence problem of $qr$ maps of a given manifold into another in Section 3. From a general point of view there is very little known of this interesting problem. One also can ask the existence of $qr$ maps with respect to contact structures, for example the Heisenberg group $H_1$ (see Section 3) has naturally such a structure. It is obtained as a limiting structure of a one parameter family of the left invariant Riemannian metrics. Some simple noninjective $qr$ maps can be exhibited, but so far there is no general theory. The $qc$ case has been more or less solved by A. Koranyi and M. Reimann [KR] and P. Pansu [Pa2].

In the classical value distribution theories by Nevanlinna and Ahlfors an important part is that they give bounds on the total contribution of the ramification indices in terms of the average covering number, in addition to the bound on the defect sum. In a very general sense that part is an extension of the Hurwitz formula which relates the total ramification to the degree of a rational function. It is most likely that something similar is true for $qr$ maps, but there does not even exist a good guess for a theorem. Another question which has been open for some time is a converse relation to the defect relation. Namely, we ask how much can, say $q$ points be covered more than the average covering number. Relatively sharp results are known in the classial case, see [Mi2], [T]. For $qr$ maps in all dimensions relationships in this direction were proved in [R2] for one point, and some of the results were new for the classical case, too. But there exists so far no general statement in the spirit of Miles's paper [Mi2].

### References

[A]     L.V. Ahlfors, Zur Theorie der Überlagerungsflächen, Acta Math. 65 (1935), 157–194.

[D]     D. Drasin, The inverse problem of the Nevanlinna theory, Acta Math. 138 (1977), 83–151.

[EL]    A. Eremenko and J. Lewis, Uniform limits of certain A–harmonic functions with application to quasiregular mappings, Ann. Acad. Sci. Fenn. Ser. A I Math. (To appear)

[G]     M. Gromov, Structures métriques pour les variétés riemanniennes, Rédigé par J. Lafontaine et P. Pansu, Cedic, Paris, 1981.

[H1]    I. Holopainen, Nonlinear potential theory and quasiregular mappings on Rie-
        mannian manifolds.- Ann. Acad. Sci. Fenn. A I Math. Dissertationes 74,
        1990.

[H2]    I. Holopainen, Positive solutions of quasilinear elliptic equations on Riemannian
        manifolds. (To appear)

[HR1]   I. Holopainen and S. Rickman, Classification of Riemannian manifolds in non-
        linear potential theory. (To appear)

[HR2]   I. Holopainen and S. Rickman, A Picard type theorem for quasiregular mappings
        into manifolds with many ends. (To appear)

[HR3]   I. Holopainen and S. Rickman, Quasiregular mappings of the Heisenberg group.
        (In preparation)

[I]     T. Iwaniec, p-harmonic tensors and quasiregular mappings. (Preprint)

[IM]    T. Iwaniec and G. Martin, Quasiregular mappings in even dimensions, Acta
        Math. (To appear)

[J]     P. Järvi, On the behavior of quasiregular mappings in the neighborhood of
        an isolated singularity, Ann. Acad. Sci. Fenn. Ser. A I Math. 15 (1990),
        341–353.

[Jo]    J. Jormakka, The existence of quasiregular mappings from $R^3$ to closed ori-
        entable 3–manifolds, Ann. Acad. Sci. Fenn. Ser. A I Math. Dissertationes
        69 (1988), 1–40.

[KR]    A. Koranyi and H.M. Reimann, Quasiconformal mappings on the Heisenberg
        group, Invent. Math. 80 (1985), 309–338.

[M]     O. Martio, Partial differential equations and quasiregular mappings, this vol-
        ume.

[MRV1]  O. Martio, S. Rickman, and J. Väisälä, Definitions for quasiregular mappings,
        Ann. Acad. Sci. Fenn. Ser. A I 448 (1969), 1–40.

[MRV2]  O. Martio, S. Rickman, and J. Väisälä, Distortion and singularities of quasireg-
        ular mappings, Ann. Acad. Sci. Fenn. Ser. A I 465 (1970), 1–13.

[MRV3]  O. Martio, S. Rickman, and J. Väisälä, Topological and metric properties of
        quasiregular mappings, Ann. Acad. Sci. Fenn. Ser. A I 488 (1971), 1–31.

[MR]    P. Mattila and S. Rickman, Averages of the counting function of a quasiregular
        mapping, Acta Math. 143 (1979), 273–305.

[Mi1]   J. Miles, A note on Ahlfors' theory of covering surfaces, Proc. Amer. Math. Soc.
        21 (1969), 30–32.

[Mi2]   J. Miles, On the counting function for the a-values of a meromorphic function,
        Trans. Amer. Math. Soc. 147 (1970), 203–222.

[N1]    R. Nevanlinna, Zur Theorie der meromorphen Funktionen, Acta Math. 46 (1925),
        1–99.

[N2]    R. Nevanlinna, Analytic Functions, Die Grundlehren der math. Wissenschaften
        Vol. 162, Springer-Verlag, Berlin–Heidelberg–New York, 1970.

[Pa1]   P. Pansu, An isoperimetric inequality on the Heisenberg group, Proceedings of
        "Differential Geometry on Homogeneous Spaces", Torino, 1983, 159-174.

[Pa2]   P. Pansu, Quasiisométries des variétés a courbure négative, Thesis, Universite
        Paris VII, 1987.

[Pel]   K. Peltonen, On the existence of quasiregular mappings. (To appear)

[Pes] M. Pesonen, A path family approach to Ahlfors's value distribution theory, Ann. Acad. Sci. Fenn. Ser. A I Math. Dissertationes 39 (1982), 1–32.

[P] E.A. Poletskiĭ, The modulus method for non–homeomorphic quasiconformal mappings, (Russian), Mat. Sb. 83 (1970), 261–272.

[Re] Yu.G. Reshetnyak, Spatial mappings with bounded distortion, (Russian), Izdat. "Nauka", Sibirsk. Otdelenie, Novosibirsk, 1982.

[R1] S. Rickman, A quasimeromorphic mapping with given deficiencies in dimension three, Symposia Mathematica Vol. XVIII (1976), INDAM, Academic Press, London, 535–549.

[R2] S. Rickman, On the value distribution of quasimeromorphic maps, Ann. Acad. Sci. Fenn. Ser. A I Math. 2 (1976), 447–466.

[R3] S. Rickman, Asymptotic values and angular limits of quasiregular mappings of a ball, Ann. Acad. Sci. Fenn. Ser. A I Math. 5 (1980), 185–196.

[R4] S. Rickman, On the number of omitted values of entire quasiregular mappings, J. Analyse Math. 37 (1980), 100–117.

[R5] S. Rickman, A defect relation for quasimeromorphic mappings, Ann. of Math. 114 (1981), 165–191.

[R6] S. Rickman, Value distribution of quasiregular mappings, Proc. Value Distribution Theory, Joensuu 1981, Lecture Notes in Math. Vol. 981, 220–245, Springer-Verlag, Berlin–Heidelberg–New York, 1983.

[R7] S. Rickman, Quasiregular mappings and metrics on the $n$–sphere with punctures, Comment. Math. Helv. 59 (1984), 134–148.

[R8] S. Rickman, The analogue of Picard's theorem for quasiregular mappings in dimension three, Acta Math. 154 (1985), 195–242.

[R9] S. Rickman, Existence of quasiregular mappings, Proceedings of the Workshop on Holomorphic Functions and Moduli, MSRI, Berkeley, Springer–Verlag (1988), 179–185.

[R10] S. Rickman, Topics in the theory of quasiregular mappings, Aspects of Mathematics, Vol. E 12, Friedr. Vieweg & Sohn (1988), 147–189.

[R11] S. Rickman, Defect relation and its realization for quasiregular mappings. (In preparation)

[R12] S. Rickman, Quasiregular Mappings. (In preparation)

[T] S. Toppila, On the counting function for the $a$–values of a meromorphic function, Ann. Acad. Sci. Fenn. Ser. A I Math. 2 (1976), 565–572.

[V1] J. Väisälä, Lectures on $n$–Dimensional Quasiconformal Mappings, Lecture Notes in Math. Vol. 229, Springer-Verlag, Berlin– Heidelberg–New York, 1971.

[V2] J. Väisälä, Modulus and capacity inequalities for quasiregular mappings, Ann. Acad. Sci. Fenn. Ser. A I 509 (1972), 1–14.

[Vu1] M. Vuorinen, Conformal Geometry and Quasiregular Mappings, Lecture Notes in Math. Vol. 1319, Springer–Verlag, Berlin–Heidelberg, 1988.

[Vu2] M. Vuorinen, Picard's theorem for entire quasiregular mappings, Proc. Amer. Math. Soc. 107 (1989), 383–394.

[Z] V.A. Zorich, The theorem of M.A. Lavrent'ev on quasiconformal mappings in space, (Russian), Mat. Sb. 74 (1967), 417–433.

Quasiconformal Space Mappings
– A collection of surveys 1960–1990
Springer–Verlag (1992), 104–118
Lecture Notes in Mathematics Vol. 1508

# TOPOLOGICAL PROPERTIES OF QUASIREGULAR MAPPINGS

Uri Srebro

Technion, Haifa, Israel

## 0. INTRODUCTION

The similarity between quasiregular (qr) mappings and holomorphic functions of one variable is clearest in their topological, metric, and measure-theoretic properties. These are also areas in which one encounters the most interesting and striking differences between the theories of two dimensions and higher dimensions.

We shall describe here only some of the topological properties of quasiregular mappings. Value distribution theory and removability theorems, for instance, are hardly mentioned here. A survey of the first topic can be found in Rickman's survey in this volume. Important developments in the second topic have been achieved recently and can be found in [I], [IM], [JV], [KM], [RI7].

In the following list are some of the topics and questions to be discussed here:
1. What is the local structure of quasiregular mappings?
2. How large can the local topological index of a quasiregular mapping be?
3. Local and global injectivity properties of quasiregular mappings.
4. Entire functions, isolated singularities, and boundary behavior.
5. Periodic and automorphic quasiregular mappings.

Several open problems will also be mentioned. Additional open problems can be found in [V3] and [VU2, p. 193].

Hereafter, whenever we write $f: D \to R^n$ we assume that $D$ is a domain in $R^n$, that $n \geq 2$, and that $f$ is continuous.

**Definition:** A function $f: D \to R^n$ is *quasiregular* if $f \in W^1_{n,\text{loc}}$ and if there exists a constant $K$, $1 \leq K < \infty$, such that $|f'(x)|^n \leq K J(x, f)$ a.e. in $D$, where

$f'$ is the formal derivative and $J(x, f) = \det f'(x)$ is the Jacobian of $f$ at $x$. A homeomorphic quasiregular mapping is called *quasiconformal*.

For alternative definitions and general information on quasiregular mappings see the books by Vuorinen 1988 [VU2], Reshetnyak 1989 [R2] and Rickman (to appear) [RI6], and the survey article by Väisälä 1978 [V3].

# 1. WHAT IS THE LOCAL STRUCTURE OF QUASIREGULAR MAPPINGS?

The following theorem describes some of the basic topological properties which quasiregular mappings and holomorphic functions of one variable have in common.

**1.1 Theorem** (Reshetnyak 1966 [R1]): If $f: D \to R^n$ is a nonconstant quasiregular mapping, then
(i) $f$ is open, i.e. maps open sets onto open sets,
(ii) $f$ is discrete, i.e. $f^{-1}(y)$ is discrete in $D$ for all $y$ in $R^n$,
(iii) $f$ is sense-preserving.

A complete understanding of the local structure of open discrete mappings and of quasiregular mappings in the plane is provided by the following generalized form of Stoïlow's Theorem :

**1.2 Theorem** (Stoïlow 1938 [ST]): If $f: D \to R^2$ is open discrete (or nonconstant and quasiregular), then $f$ has the following decomposition: $f = g \circ h$, where $h$ is a homeomorphism (or a quasiconformal mapping) of $D$ into $R^2$ and $g: fD \to C$ is a nonconstant holomorphic mapping.

**1.3** As a corollary of 1.2 we can conclude that if $f: D \to R^2$ is open discrete (or nonconstant and quasiregular) and $x_o$ is a point in $D$, then
(i) at $x_o$, $f$ is locally topologically (or quasiconformally) equivalent to the mapping $z^d$, $z \in C$, for some integer $d = d(x_o) \geq 1$,
(ii) the set of points of $D$ where $f$ is not a local homeomorphism is discrete in $D$.

**1.4 The branch set** (sometimes called the critical set) $B_f$ of a map $f: D \to R^n$ is defined as the set of points of $D$ where $f$ is not a local homeomorphism.

It is clear that the branch set is closed in $D$ for any $n \geq 2$, and that for an open discrete map $f: D \to R^2$ the branch set is either empty or a discrete set in $D$. Indeed, by Stoïlow's Theorem, $f = g \circ h$ for some homeomorphism $h$ and nonconstant analytic function $g$. The branch set of $g$ is $\{w: g'(w) = 0\}$, which is either empty or discrete in $hD$. Hence $B_f = h^{-1}B_g$ is either empty or discrete in $D$.

It is not hard to see that in all higher dimensions $n \geq 3$, the branch set never contains isolated points. The following theorem suggests that the dimension of the branch set of an open discrete map cannot be too large either.

**1.5 Theorem** (Chernavskiĭ 1965 [CHE1], [CHE2]; see also Väisälä 1966 [V1]): If $f: D \to R^n$, $n \geq 2$, is open discrete, then $\dim B_f = \dim f B_f \leq n - 2$.

Here dim is the topological dimension. The inequalities in this theorem show that neither $B_f$ nor $fB_f$ can separate any subdomain of $R^n$.

## 2. MORE ON THE LOCAL STRUCTURE—EXAMPLES

The mappings in the following examples are either quasiregular or can be made so by some modification. The examples are topologically typical in various situations which will be described in Section 3.

The simplest example of a quasiregular mapping in $R^n$, $n \geq 3$, which is not a local homeomorphism is the following winding map.

**2.1 The $k$-winding map** $f_k: R^n \to R^n$, $k, n \geq 2$, is defined by

$$(r, \theta, y) \mapsto (r, k\theta, y),$$

where $r$ and $\theta$ are polar coordinates in the $(x_1, x_2)$-plane and $y = (x_3, \ldots, x_n)$.

Note that for $n = 2$ the mapping $f_k$ is topologically equivalent to $z^k$, $z \in C$. In higher dimensions $f_k$ is topologically equivalent to $z^k \times id$. The branch set of $f_k$ and its image are the $(n-2)$-dimensional subspace $\{x \in R^n: x_1 = x_2 = 0\}$. The map $f_k$ is sense-preserving. It is locally injective off the branch set and locally $k$-to-1 at every point of the branch set, i.e. the local topological index $i(x, f_k)$ is $k$ at every point $x$ of the branch set and 1 elsewhere.

**The local topological index** of a sense-preserving map $f: D \to R^n$ can be defined by

$$i(x, f) = \lim_{r \to 0} \sup_y \text{card} \left( f^{-1}(y) \cap B^n(x, r) \right).$$

Calculations show that the mappings $f_k$ are quasiregular with inner dilatation $K_I(f_k) = k$ in all dimensions $n \geq 2$. In particular, for the 2-winding map $K_I(f_2) = 2$. In view of this simple example and the fact that holomorphic functions have inner dilatation $K_I = 1$ and may have arbitrarily large local index the following conjecture is an interesting challenge.

**2.2 Conjecture** (Martio 1970 [M1]): If $f: D \to R^n$ is quasiregular with $K_I(f) < 2$ and if $n \geq 3$, then $f$ is a local homeomorphism.

Note that in dimensions $n \geq 3$ locally homeomorphic quasiregular mappings have quite strong rigidity properties, as shown below in Section 5. The following known facts make this conjecture even more interesting.

**2.3 Known facts** related to conjecture 2.2
(i) Martio 1970 [M1]: Let $f: D \to R^n$ be quasiregular, where $n \geq 3$. If $B_f$ includes a rectifiable arc then $K_I(f) \geq 2$.
(ii) Martio, Rickman, and Väisälä 1971 [MRV3], Gol'dshtein 1971 [G]: There exists a universal constant $K > 1$ such that every quasiregular mapping $f: D \to R^n$, $n \geq 3$, with $K_O(f) < K$ is a local homeomorphism.

**2.4 The cone of rational functions** (Martio and Srebro 1979 [MS5]). Let $g: S^2 \to S^2$ be a rational function, i.e. a quotient of two polynomials realized as a self-map of the Riemann sphere. The *cone* or *radial extension* of $g$ is the mapping $f: B^3 \to B^3$ defined by $f(tx) = tg(x)$ for $x \in S^2$ and $0 \le t \le 1$.

Note that

1. $f$ is continuous, open, discrete and can be made quasiregular or piecewise linear by a suitable modification,
2. $i(0, f) = \deg f = \deg g$,
3. $B_f$ and $fB_f$ have a ray structure at 0, i.e. each is a union of rays emanating from 0.

Such maps can be used (cf. Rickman 1969 [RI1]) to show that the local index of a quasiregular mapping $f: D \to R^n$ can be arbitrarily large at an isolated point even if its dilatation is bounded above by some universal constant $M$.

**2.5 A mapping associated with a path of rational functions** (Martio and Srebro 1979 [MS5]). Let $M_k$ denote the space of all rational functions of degree $k$ acting on $S^2$. Given a path $\alpha: [0, 1] \to M_k$, the mapping $f_\alpha: B^3 \to B^3$ associated with $\alpha$ is defined by $f_\alpha(tx) = tf(x)$ for $x \in S^2$ and $0 \le t < 1$.

Note that a cone of a rational function is a mapping associated with a degenerate path $\alpha =$ const. and that

1. $f_\alpha$ is continuous, open, discrete,
2. $i(0, f_\alpha) = k$,
3. $f_\alpha^{-1}\left(S^2(t)\right) = S^2(t)$ and $f_\alpha^{-1}\left(B^3(t)\right) = B^3(t)$, $0 < t < 1$.

**2.6** An open discrete mapping, or even a quasiregular map $f: D \to R^3$, need not be topologically equivalent to one of the mappings presented above. Church and Hemmingsen 1960 [CH] constructed an open discrete map $f: R^3 \to R^3$ with $\deg f = 3$ such that the origin is a point component of $B_f$ and also of $fB_f$. In their example $B_f$ is an infinite chain of linked circles accumulating at the origin. Note that in the examples in 2.4 and 2.5 $B_f$ is connected and $B_f$ is locally connected at the origin, which is not the case in Church and Hemmingsen's example.

The following theorems characterize the maps that are topologically equivalent to one of the examples in 2.1, 2.4, and 2.5.

**2.7 Theorem** (Martio, Rickman, and Väisälä, 1971 [MRV3]): Let $f: D \to R^n$ be open discrete (or quasiregular) and $x_o$ a point in $D$. If there exist a neighborhood $V$ of $f(x_o)$ and a homeomorphism $\varphi: V \to R^n$ such that

$$\varphi(V \cap fB_f) \subset R^{n-2} \subset R^n,$$

then at $x_o$, locally, $f$ is topologically (or quasiconformally) equivalent to a winding mapping.

**2.8 Theorem** (Martio and Srebro 1979 [MS5]): Let $f: D \to R^3$ be open discrete (or quasiregular) and $x_o$ a point in $B_f$. If there exists a neighborhood $V$ of $f(x_o)$ such that $V \cap fB_f$ has a ray structure, then at $x_o$, locally, $f$ is topologically (or quasiconformally) equivalent to a cone of a rational function.

**2.9 Theorem** (Martio and Srebro 1979 [MS5]): Let $f: D \to R^3$ be open discrete (or quasiregular) and $x_o$ a point in $B_f$. If there exists a homeomorphism $\varphi$ of $B^3$ onto a neighborhood of $f(x_0)$ such that $\varphi(0) = f(x_o)$ and such that the boundary of the $x_o$-component of $f^{-1}\varphi\left(B^3(t)\right)$ is a connected 2-manifold for all $t \in (0,1)$, then at $x_o$, locally, $f$ is topologically (or quasiconformally) equivalent to a map associated with a path of rational functions.

## 3. MORE ON THE BRANCH SET

Recall from Theorem 1.5 that $\dim B_f = \dim f B_f \leq n - 2$ for all open discrete maps $f: D \to R^n$, $n \geq 2$. In the examples of Section 2 the branch set and its image were of co-dimension two. We are thus led to the following questions.

### 3.1 Questions
(i) Can an open discrete map $f: D \to R^n$, $n \geq 3$, have a nonempty branch set $B_f$ with $\dim B_f < n - 2$?
(ii) As in (i) but for a quasiregular mapping.

**3.2** The first question was answered affirmatively for $n \geq 5$ by Church and Timorian [CT], who constructed for each $n \geq 5$ an open discrete map $f: S^n \to S^n$ such that $B_f$ and $f B_f$ are homeomorphic to $S^{n-4}$, i.e. $\dim B_f = \dim f B_f = n - 4$.

The idea behind the construction was to start with the covering of the Poincaré homology sphere $X$ by $S^3$ and use the fact that the double suspension of each of these spaces is homeomorphic to $S^5$. The natural extension of the covering map yielded the desired map for $n = 5$, and this map can be extended to all $S^n, n > 5$.

Question 3.1(i) remais open for $n = 3$ and $n = 4$. Question 3.1(ii) is still open for all $n > 2$. We rephrase it and add as 3.3 (iii) another problem that was found by Vuorinen in 1976 (cf. [VU2, 9.18 and p. 193(4)]. ) The problem 3.3 (iii) is still open, although unpublished partial results exist.

### 3.3 Problems
(i) Can an open discrete map $f: D \to R^n$, $n = 3, 4$, have a nonempty branch set $B_f$ with $\dim B_f < n - 2$?
(ii) Is there a quasiregular map $f: D \to R^n$ with $B_f \neq \emptyset$ and with $\dim B_f$ less than $n - 2$? In particular, can the Church and Timorian example be made quasiregular by a suitable modification?
(iii) Can a noninjective proper open discrete map $f: B^n \to R^n$ have a compact branch set if $n \geq 3$?

The following known facts indicate that the second problem does not have a simple answer.

From now on we shall use $\mathcal{H}^k$ to denote the $k$-dimensional Hausdorff measure.

### 3.4 Facts
(i) Martio, Rickman, and Väisälä 1971 [MRV3]; Gol'dshtein 1971 [G]: If $f: D \to R^n$ is open discrete with $B_f \neq \emptyset$ then $\mathcal{H}^{n-2}(f B_f) > 0$.

(ii) Martio and Rickman 1971 [MR2]: If $f$ is as in (i) and $n = 3$, then $\mathcal{H}^1(fB_f) > 0$.

## 4. HOW LARGE CAN THE LOCAL INDEX BE?

**4.1 In two dimensions**, we may confine ourselves to analytic functions, thereby concluding that plane quasiregular mappings may have arbitrarily large local index on any given discrete set which does not accumulate in the domain of the mapping.

**4.2 In higher dimensions** $n \geq 3$ we have the following three results:
(i) The local index can be arbitrarily large at a point (Rickman 1969 [RI1]), i.e. there exists $K < \infty$ such that for every positive integer $k$ there exists a quasiregular map $f: B^n \to B^n$ with $K(f) < K$ and $i(0, f) > k$.
(ii) The local index cannot be uniformly too large on a continuum (Martio 1970 [M1]) or on a sequence of points which converges too slowly to a point of the domain (Rickman and Srebro 1986 [RS]).
(iii) There exists a constant $K < \infty$ such that for every positive integer $k$ there exists a quasimeromorphic mapping $f: R^3 \to \bar{R}^3$ with $K(f) < K$ such that $\{x \in R^3 : i(x, f) \geq k\}$ is a Cantor set (Rickman 1985 [RI5]).

## 5. LOCAL AND GLOBAL INJECTIVITY OF QUASIREGULAR MAPS

The most striking differences between two-dimensional quasiregular mappings and those in higher dimensions occur in questions of local and global injectivity and in the existence of a universal radius of injectivity. As pointed out earlier, contrary to the situation in the plane, a quasiregular mapping $f: D \to R^n$ with small dilatation is locally injective if $n \geq 3$. It can be shown that $f: B^n \to R^n$ is (globally) injective if $n \geq 3$ and if the dilatation of $f$ is sufficiently small [SA]. For the same result in domains other than $B^n$ see [MSA].

Other results in this direction are presented in the following theorems.

**5.1 Theorem** (Zorich 1967 [ZO1]): If $n \geq 3$ and $f: R^n \to R^n$ is quasiregular and locally injective, then $f$ is injective.

A related result is the following removability theorem.

**5.2 Theorem** (Agard and Marden 1970 [AM], Zorich 1970 [ZO2]): If $n \geq 3$ and $f: B^n \setminus \{0\} \to R^n$ is quasiregular and locally injective, then $f$ is injective and $f$ is a restriction of a quasiconformal map of $B^n$.

The local version of Zorich's Theorem is given in the following theorem.

**5.3 Theorem** (Martio, Rickman, and Väisälä 1971 [MRV3]): For every integer $n \geq 3$ and every $K$ in $(1, \infty)$ there exists a number $r = r(n, K)$ in $(0, 1)$ such that if $f: B^n \to R^n$ is $K$-quasiregular and locally injective then $f$ is injective in $B^n(r)$.

**5.4 Remarks**

(i) The three theorems 5.1–5.3 are true also for quasimeromorphic mappings in $R^n$, $n \geq 3$, [MS4].

(ii) The three results 5.1–5.3 and their extensions to quasimeromorphic mappings are false in two dimensions. Counterexamples can easily be constructed from polynomial and exponential functions.

(iii) 5.3 implies 5.1.

## 6. ENTIRE FUNCTIONS, ISOLATED SINGULARITIES, AND BOUNDARY BEHAVIOR

**6.1 Entire functions.** The notion *entire* will be used for mappings which are quasiregular in the whole space $R^n$, $n > 3$. The following theorem holds in all dimensions and applies to all entire functions for which $\infty$ is a removable singularity.

**Theorem** (cf. [MRV2], [MS1]) : Let $f : R^n \rightarrow R^n$, $n \geq 2$, be a quasiregular mapping.

(i) Then $f(x) \rightarrow \infty$ as $x \rightarrow \infty$ if and only if $\{\text{card } f^{-1}(y)) : y \in R^n\}$ is a bounded subset of $R$.

(ii) If $f(x) \rightarrow \infty$ as $x \rightarrow \infty$ then $f$ extends to a quasiregular map $F : \bar{R}^n \rightarrow \bar{R}^n$ with $F(\infty) = \infty$, with $f(R^n) = F(R^n) = R^n$, and with

$$\sum_{x \in f^{-1}(y)} i(x, f) = \deg F < \infty \text{ for all } y \in R^n.$$

**6.2** We now turn to the case where $\infty$ is an essential, i.e. nonremovable, singular point.

For some time it was not known to what extent Picard's Theorem on exceptional sets holds in higher dimensions. It was clear that Stoïlow's Theorem could be used to extend Picard's Theorem to plane quasiregular mappings. An example by Zorich (see 6.8 below) showed that entire quasiregular mappings may omit a point in $R^n$ for any $n \geq 2$. On the other hand, it is relatively easy to show that Liouville's Theorem on bounded entire functions holds for quasiregular mappings in all dimensions (cf. Martio, Rickman, and Väisälä 1969, 1970 [MRV1-2]). Even the stronger result, which says that a nonconstant entire function cannot omit a set of positive capacity and in particular cannot omit a nondegenerate continuum (cf. [MRV2]), is true.

It is also known that Iversen's Theorem, which says that every value omitted in a neighborhood of an isolated essential singular point is an asymptotic value, holds for quasiregular mappings in all dimensions.

In the early eighties Rickman showed that a nonconstant quasiregular mapping $f$ defined in $R^n$ can omit at most a finite number of points, where the cardinality of the omitted set depends only on the dimension $n$ and the dilatation of $f$.

**6.3 Theorem** (Rickman 1980 [RI2]): For every integer $n \geq 2$ and every constant $K$ in $(1, \infty)$ there is an integer $q = q(n, K)$ such that if $f: R^n \to R^n$ is nonconstant $K$-quasiregular, then card $(R^n \setminus f(R^n)) < q$.

This theorem left open the problem of how large the omitted set $\bar{R}^n \setminus f(R^n)$ can be in a given dimension $n \geq 3$.

For $n = 3$ this question was settled by Rickman [RI4] in the following theorem, which shows that for $n = 3$ the cardinality of the omitted set can be arbitrarily large.

**6.4 Theorem** (Rickman 1985 [RI4]): For each $p > 0$ there is a quasiregular map $f: R^3 \to R^3$ such that card$(R^3 \setminus f(R^3)) = p$.

Theorems 6.1, 6.3, and 6.4 are part of an extensive study of value distribution theory of quasiregular mappings. This study includes an extension of Nevanlinna's Theory (see Rickman's survey in this volume).

Several recent results about isolated singularities of quasiregular mappings are given in [J] and [VU3].

**6.5 Isolated singular points.** Theorems 6.1, 6.3, and 6.4 above, have natural extensions to theorems on isolated singular points of quasiregular mappings in an arbitrary domain.

**6.6 Boundary behavior.** Much of the theory carries over from the plane to higher dimensions, but there are also surprising differences. For example, Lindelöf's Theorem on the uniqueness of an asymptotic value of a bounded holomorphic function in a disk [N] does not hold for quasiregular mappings in higher dimensions. For a survey of this topic see [VU1], [VU2, Ch. 15].

**Theorem** (Rickman [RI3]): For each $n \geq 3$ there exists a bounded quasiregular mapping $f: B^n \to R^n$ which has infinitely many asymptotic values at some boundary point.

**6.7 Fatou's Theorem** for bounded holomorphic functions in a disk asserts the existence of a radial limit at almost every boundary point [N, p. 198]. The theorem is false for plane quasiregular mappings. A counterexample is $g \circ h$, where $h$ is a quasiconformal automorphism of $B^2$ whose boundary extension takes a subset $E$ of $\partial B^2$ with a positive linear measure onto a set of zero measure in $\partial B^2$, and where $g$ is a holomorphic function which has no radial limits on $hE$.

It is not known whether Fatou's Theorem holds for bounded quasiregular mappings in higher dimensions. Even the following problem is still open.

**Problem.** Let $f: B^3 \to B^3$ be a nonconstant quasiregular mapping. Is it true that $f$ must have a radial limit at some boundary point?

**6.8 Zorich's function.** We close this section with the construction of Zorich's function $f: R^n \to R^n \setminus \{0\}$. The explicit construction will be given only for $n = 3$. The extension to all $n > 3$ is straightforward.

Zorich's function plays an important role as a building block in many examples in value distribution theory and in boundary behavior problems.

**Step I**: Consider the infinite square cylinder

$$P_1 = \{x \in R^3 : 0 < x_1 < 1, \quad 0 < x_2 < 1, \quad -\infty < x_3 < \infty\}.$$

There is an explicit homeomorphism $f$ of $\bar{P}_1$ into the closed half space

$$\bar{H}^3 = \{x \in R^3 : x_3 \geq 0\}$$

such that $f|P_1$ is quasiconformal,

$$fP_1 = H^3, \qquad \lim_{x_3 \to -\infty} f(x) = 0, \qquad \lim_{x_3 \to \infty} f(x) = \infty,$$

and such that the infinite edges $\ell_1, \ell_2, \ell_3$ and $\ell_4$ are mapped onto four rays $\ell_1' = -\ell_3'$ and $\ell_2' = -\ell_4'$, $\ell_1' \perp \ell_2'$, in $\partial H^3$ emanating from the origin (cf.[ZO1],[MS1]).

**Step II**: Extend $f$ by reflections in the faces of $P_1$ and their images in $\partial H^3$, respectively. This extended map, also denoted by $f$, is quasiregular by the reflection principle for quasiregular mappings.

**Step III**: Continue and extend $f$ by all possible reflections.

This last step leads to a quasiregular mapping, still denoted by $f$, which maps $R^3$ onto $R^3 \setminus \{0\}$ and has the following properties:
(i) The branch $B_f$ is the union of the edges $\ell_i$, $i = 1, \ldots, 4$, and their images under the translations $x \mapsto x + ke_1 + le_2$, $k, l \in Z$, while

$$fB_f = \bigcup_{i=1}^{4} \ell_i'.$$

(ii) $f$ is 2-periodic with fundamental periods $2e_1$ and $2e_2$, i.e. $f(x + 2ke_i) = f(x)$ for $i = 1, 2$ and $k \in Z$, i.e. $f$ is automorphic (that is invariant) for the group $G$ which is generated by the translations $x \mapsto x + 2e_i$, $i = 1, 2$.
(iii) $f$ induces a quasiregular mapping $\tilde{f}: R^3/G \to R^3 \setminus \{0\}$ with $\deg(\tilde{f}) = 2$.

## 7. PERIODIC QUASIMEROMORPHIC MAPPINGS

A nonconstant mapping $f: R^n \to \bar{R}^n$ is *periodic* if there exists $a \in R^n \setminus \{0\}$ such that $f(x + a) = f(x)$ for all $x \in R^n$. If $f$ is open, discrete, and periodic, then the set $M$ of its periods is a finite-dimensional module with $1 \leq k = \dim M \leq n$. We then say that $f$ is *k-periodic*.

A basis $\{a_1, \ldots, a_k\}$ of the period module $M$ of a $k$-periodic mapping can be completed to a basis $\{a_1, \ldots, a_n\}$ of $R^n$. Then

$$F = \{x = \sum_{i=1}^{n} t_i a_i \ : \ 0 \le t_i < 1 \text{ if } 1 \le i \le k \text{ and } t_i \in R \text{ if } k+1 \le i \le n\}$$

is a *fundamental set* for $f$ and for the group $G$ generated by the translations $x \mapsto x + a_i$, $i = 1, \ldots, k$, while $D = \text{int } F$ is a *fundamental domain* for $f$ and $G$. Note that $f$ induces a quasiregular map $\tilde{f}: R^n/G \to \bar{R}^n$ with $f = \tilde{f} \circ \pi$, where $\pi: R^n \to R^n/G$ is the natural projection.

**7.1 $n$-periodic quasimeromorphic mappings in $R^n$.** One can obtain an example of such a map by first mapping the cube $Q = \{x \in R^n : 0 < x_i < 1, \ i = 1, \ldots, n\}$ quasiconformally onto $H^n$ and then extending by all possible reflections, as in the construction of Zorich's Function in 6.8. The result is a quasimeromorphic mapping of $R^n$ onto $\bar{R}^n$ with periods $2e_1, \ldots, 2e_n$ and a fundamental set $F = \{x \in R^n : -1 \le x_i < 1, \ 1 \le i \le n\}$. Note that $(F, f)$ covers every point of $\bar{R}^n$ exactly $2^{n-1}$ times, counting multiplicities. This situation is quite typical, as the following theorem indicates.

**7.2 Theorem** (Martio and Srebro 1975 [MS1]): If $f: R^n \to \bar{R}^n$ is $n$-periodic open discrete with a fundamental set $F$, then $f(R^n) = f(F) = \bar{R}^n$, and every point of $\bar{R}^n$ is covered by $F$ the same (finite) number of times, counting multiplicities, i.e. there is an integer $d > 0$ such that

$$\sum_{x \in f^{-1}(y) \cap F} i(x, f) = d \ \text{ for all } y \in \bar{R}^n.$$

**7.3 $(n-1)$-periodic quasimeromorphic mappings in $R^n$.** The exponential function is an example of such a function in the plane, and Zorich's function is an example of such a mapping in higher-dimensional space.

Some of the known topological properties of $(n-1)$-periodic quasiregular mappings are summarized in the following theorem.

**7.4 Theorem** (Martio and Srebro 1975 [MS1]): Let $f: R^n \to \bar{R}^n$ be an $(n-1)$-periodic quasimeromorphic mapping with periods $a_1, \ldots, a_{n-1}$, all perpendicular to $e_n$, and let $F$ be a fundamental set determined by $a_1, \ldots a_{n-1}$. Then
(i) The two limits

$$\lim f(x) \text{ as } x_n \to +\infty \text{ and } x \in F$$

and

$$\lim f(x) \text{ as } x_n \to -\infty \text{ and } x \in F$$

exist in $\bar{R}^n$ if and only if card $(f^{-1}(y) \cap F)$ is finite and bounded,
(ii) If card $(f^{-1}(y) \cap F)$ is bounded, then card $(R^n \setminus f(R^n)) \le 2$,
(iii) If $n > 2$, then $B_f \ne \emptyset$.

An interesting phenomenon occurs when $1 \le k \le n-2$:

**7.5 Theorem** (Martio 1976 [M2]): Let $1 \leq k \leq n - 2$ and let $f: R^n \to \bar{R}^n$, $n \geq 3$, be a $k$-periodic quasimeromorphic mapping with fundamental set $F$. Then card $(f^{-1}(y) \cap F) = \infty$ for almost all $y$ in $\bar{R}^n$.

## 8. AUTOMORPHIC QUASIMEROMORPHIC MAPPINGS

Let $G$ be a discontinuous Möbius group acting on $\bar{R}^n$, and $D \subset R^n$ an invariant subdomain of the regular set of $G$; that is, $D$ is a domain such that $gD = D$ for all $g$ in $G$ and such that no point of $D$ is a limit point of $G$. An open discrete mapping $f: D \to \bar{R}^n$ is *automorphic* for $G$ if $f \circ g = f$ for all $g$ in $G$. Such a map induces an open discrete map $\tilde{f}: D/G \to \bar{R}^n$ such that $f = \pi \circ \tilde{f}$, where $\pi: D \to D/G$ is the natural projection.

**8.1 Existence problem.** Given a discrete group $G$ acting on a domain $D$, does there exist a nonconstant meromorphic map $f$ on $D$ which is automorphic with respect to $G$? The next two theorems give a partial solution to this problem.

**8.2 Theorem** (Martio and Srebro 1977 [MS3]): Let $G$ be a discrete Möbius group acting on $B^n$. If $B^n/G$ is of finite hyperbolic volume then there exists a nonconstant quasimeromorphic mapping $f: B^n \to \bar{R}^n$ which is automorphic with respect to $G$.

**8.3 Theorem:** Let $G$ be a discrete Möbius group acting on a domain $D$. If no element of $G$ has a fixed point in $D$, then there exists a nonconstant quasimeromorphic mapping which is automorphic with respect to $G$.

If $G$ has no fixed point in $D$, then $D/G$ is a manifold which carries either the natural euclidean or hyperbolic structure, and the natural projection $\pi: D \to D/G$ is conformal. A theorem of Peltonen 1988 [P] asserts the existence of a quasiregular map $g: D/G \to \bar{R}^n$, and consequently $g \circ \pi$ is quasimeromorphic and automorphic with respect to $G$.

A special case of the last theorem for $D = B^n$ was proved by Tukia [T].

It can be shown that there are discrete Möbius groups acting on $B^3$ which do not carry quasimeromorphic automorphic mappings.

**8.4 Local injectivity of automorphic mappings.** There are also differences between the local injectivity properties of plane automorphic mappings and automorphic mappings in higher dimensions. The following theorem illustrates some of these differences.

**Theorem** (Martio and Srebro [MS6]): Let $f: B^n \to \bar{R}^n$ be a quasimeromorphic mapping which is automorphic with respect to a discrete Möbius group $G$ that acts on $B^n$. If $n \geq 3$ then any of the following conditions implies that $f$ has a nonempty branch set:
(i) $G$ has parabolic elements.
(ii) $B^n/G$ is of finite hyperbolic volume.

(iii) $G$ has loxodromic elements of arbitrarily small translation length.

An element $g$ of $G$ is *parabolic* if it has a unique fixed point, and it is *loxodromic* if it has exactly two fixed points in $\partial B^n$ and none in $B^n$. The *translation length* of a loxodromic element $g$ in $G$ is sup $\rho(x, g(x))$, where $\rho$ is hyperbolic distance in $B^n$ and the supremum is taken over all $x \in B^n$.

**8.5 Remarks.** (1) The elliptic modular function is a two-dimensional counterexample to (i) and (ii) of Theorem 8.4.

(2) To show that (iii) is not true in the plane, we now construct a discrete Möbius group $G$ acting on $H^2 = \{z \in C: \Im z > 0\}$ with loxodromic elements of arbitrarily small translation length, and a locally univalent function $f$ in $H^2$ which is automorphic with respect to $G$.

**Example** [MS6]. Choose two increasing sequences of positive real numbers $x_n$ and $r_n$, $n = 1, 2, \ldots$, such that the closed disks $D_n = \{z : |z - x_n| \leq r_n\}$ are disjoint and such that the hyperbolic distance $d_n$ between $D_n \cap H^2$ and the imaginary axis $l$ tends to zero as $n \to \infty$. Let $T_n$ be the reflection in $\partial D_n$ followed by a reflection in $l$, and $G$ the group generated by all $T_n$, $n = 1, 2, \ldots$. Then $G$ is a discrete Möbius group acting on $H^2$, with loxodromic elements of arbitrarily small translation length. We now construct a locally univalent holomorphic function $f$ in $H^2$ which is automorphic with respect to $G$.

Let $D$ denote the sub-domain of $H^2$ which is exterior to all disks $D_n$ and to all disks $T_n(D_n)$. Then $D$ is simply-connected and it is a fundamental domain for $G$. Let $h$ be a homeomorphism of $\bar{D}$ onto $\bar{H}^2$ such that $h(0) = 0$, $h(\infty) = \infty$, $h(i) = i$ and such that $h$ is conformal in $D$. Then $h$ is symmetric with respect to the imaginary axis $l$, and therefore the function $f$ defined by $f(z) = (h(z))^2$ agrees on points of $\partial D$ which are equivalent under $G$. Also $f(z)$ is conformal in $D$. Since the elements of $G$ are conformal and since $G\bar{D}$ is a tiling of $H^2$ the mapping $f$ can be extended holomorphically to $H^2$ if one defines $f(z) = f(g(z))$ for all $g$ in $G$. It is easy to check that $f$ is locally univalent, and it is clear from the definition that $f$ is automorphic with respect to $G$.

Note that the surface $H^2/G$ has closed geodesics of arbitrarily small length, and the mapping $f$ induces a conformal immersion $\tilde{f}$ of $H^2/G$ into $S^2$. However, for $n > 2$, 8.4(iii) yields the following corollary.

**8.6 Corollary** (Martio and Srebro [MS6]): Let $M$ be a hyperbolic $n$-manifold. If $M$ has closed geodesics of arbitrarily small length and if $n \geq 3$, then $M$ cannot be immersed quasiconformally in $S^n$.

**8.7 Problem.** Let $M$ be a hyperbolic $n$-manifold with $n \geq 3$. Is it true that if the length of each closed geodesic of $M$ is at least $\delta > 0$, then $M$ can be immersed quasiconformally in $S^n$?

**8.8 Boundary behavior.** Let $f$ be a nonconstant quasimeromorphic mapping which is automorphic with respect to a discrete Möbius group $G$ that acts on $B^n$, $n \geq 2$. It is clear that $f$ has no continuous extension at any limit point of $G$

and that it fails to have a radial limit at any conical limit point of $G$. A point $b$ in $\partial B^n$ is a *limit point* of $G$ if the orbit of $O$ clusters at $b$; $b$ is *a conical limit point of $G$* if the orbit of $O$ intersected with some Stolz cone at $b$ clusters at $b$. This observation can be used to attack the following two questions:

**Questions.**
(i) Can $\partial B^n$ be the natural boundary of some quasimeromorphic, quasiregular, or bounded quasiregular mapping of $B^n$?
(ii) Is some version of Fatou's Theorem on the radial limits of a bounded function true in higher dimensions (cf. Problem 6.7 above)?

Indeed, by using appropriate automorphic mappings, Martio and Srebro [MS6] were able to construct the following for all $n \geq 3$:
(1) quasiregular mappings $f: B^n \to R^n$ such that $\dim(R^n \setminus fB^n) = 1$ and such that $\partial B^n$ is the natural boundary of $f$,
(2) bounded quasiregular mappings $f: B^n \to R^n$ which have no radial limit on a subset of $\partial B^n$ of positive Hausdorff dimension.

**Acknowledgements.** This work was partially supported by the Technion Fund for the Promotion of Research. The author wishes to thank Olli Martio and Glen Anderson for many valuable remarks and comments.

## References

[AM]   S. AGARD and A. MARDEN: A removable singularity theorem for local homeomorphisms, *Indiana Math. J.* **20** (1970), 455–461.

[CHE1] A. V. CHERNAVSKIĬ: Finite–to–one open mappings of manifolds. (Russian), *Mat. Sb.* **65** (1964), 357–369.

[CHE2] A. V. CHERNAVSKIĬ: Remarks on the paper "On finite–to–one mappings of manifolds", (Russian), *Mat. Sb.* **66** (1965), 471–472.

[CH]   P. T. CHURCH and E. HEMMINGSEN: Light open maps on $n$-manifolds, *Duke Math. J.* **27** (1960), 527–536.

[CT]   P. T. CHURCH and J. G. TIMOURIAN: Differentiable maps with small critical set or critical set image, *Indiana Univ. Math. J.* **27** (1978), 953–971.

[G]    V. M. GOL'DSHTEIN: The behavior of mappings with bounded distortion when the coefficient of distortion is close to unity, *Sibirsk. Math. Zh.* **12** (1971), 900–906.

[I]    T. IWANIEC: $p-$harmonic tensors and quasiregular mappings, *Ann. Math. (to appear)*.

[IM]   T. IWANIEC and G. MARTIN: Quasiregular mappings in even dimensions, *Acta Math. (to appear)*.

[J]    P. JÄRVI: On the behavior of quasiregular mappings in the neighborhood of an isolated singularity, *Ann. Acad. Sci. Fenn. Ser. A I Math.* **15** (1990), 341–353.

[JV]   P. JÄRVI and M. VUORINEN: Self-similar Cantor sets and quasiregular mappings, *J. Reine Angew. Math.* **424** (1992), 31–45.

[KM]    P. KOSKELA and O. MARTIO: Removability theorems for quasiregular
        mappings, *Ann. Acad. Sci. Fenn. Ser. A I Math.* **15** (1990), 381–399.
[M1]    O. MARTIO: A capacity inequality for quasiregular mappings, *Ann. Acad.
        Sci. Fenn. Ser. A I* **474** (1970), 1–18.
[M2]    O. MARTIO: On $k$–periodic quasiregular mappings in $R^n$, *Ann. Acad.
        Sci. Fenn. Ser. A I Math.* **1** (1975), 207–220.
[MR1]   O. MARTIO and S. RICKMAN: Boundary behavior of quasiregular map-
        pings, *Ann. Acad. Sci. Fenn. Ser. A I* **507** (1972), 1–17.
[MR2]   O. MARTIO and S. RICKMAN: Measure properties of the branch set and
        its image of quasiregular mappings, *Ann. Acad. Sci. Fenn. Ser. A I*
        **541** (1973), 1–16.
[MRV1]  O. MARTIO, S. RICKMAN, and J. VÄISÄLÄ: Definitions for quasireg-
        ular mappings, *Ann. Acad. Sci. Fenn. Ser. A I* **448** (1969), 1–40.
[MRV2]  O. MARTIO, S. RICKMAN, and J. VÄISÄLÄ: Distortion and singular-
        ities of quasiregular mappings, *Ann. Acad. Sci. Fenn. Ser. A I* **465**
        (1970), 1–13.
[MRV3]  O. MARTIO, S. RICKMAN, and VÄISÄLÄ: Topological and metric prop-
        erties of quasiregular mappings, *Ann. Acad. Sci. Fenn. Ser. A I* **488**
        (1971), 1–31.
[MSA]   O. MARTIO and J. SARVAS: Injectivity theorems in plane and space,
        *Ann. Acad. Sci. Fenn. Ser. A I Math.* 4 (1978/1979), 383–401.
[MS1]   O. MARTIO and U. SREBRO: Periodic quasimeromorphic mappings in
        $R^n$, *J. Analyse Math.* **28** (1975), 20–40.
[MS2]   O. MARTIO and U. SREBRO: Automorphic quasimeromorphic mappings
        in $R^n$, *Acta Math.* **135** (1975), 221-247.
[MS3]   O. MARTIO and U. SREBRO: On the existence of automorphic quasimero-
        morphic mappings in $R^n$, *Ann. Acad. Sci. Fenn. Ser. A I Math.* **3**
        (1977), 123–130.
[MS4]   O. MARTIO and U. SREBRO: Universal radius of injectivity for locally
        quasiconformal mappings, *Israel J. Math.* **29** (1978), 17-23.
[MS5]   O. MARTIO and U. SREBRO: On the local behavior of quasiregular maps
        and branched covering maps, *J.Analyse Math.* **36** (1979), 198–212.
[MS6]   O. MARTIO and U. SREBRO: *To appear.*
[N]     R. NEVANLINNA: Analytic Functions, Die Grundlehren der math. Wissen-
        schaften Vol. **162**, Springer-Verlag, Berlin–Heidelberg–New York, 1970.
[P]     K. PELTONEN: On the existence of quasiregular mappings, *Manuscript,*
        (1988).
[R1]    YU. G. RESHETNYAK: Space mappings with bounded distortion (Rus-
        sian), *Sibirsk. Mat. Zh.* **8** (1967), 626–659.
[R2]    YU. G. RESHETNYAK: Space Mappings with Bounded Distortion, Trans-
        lations of Mathematical Monographs Vol. **73**, Amer. Math. Soc.
        Providence, R.I., 1989.
[RI1]   S. RICKMAN: Quasiregular Mappings, *Proc. Romanian-Finnish Seminar
        on Teichmüller spaces and quasiconformal mappings*, Brasov, (1969),
        261–271.

[RI2]   S. RICKMAN: On the number of omitted values of entire quasiregular mappings, *J. Analyse Math.* **37** (1980), 100–117.

[RI3]   S. RICKMAN: Asymptotic values and angular limits of quasiregular mappings of a ball, *Ann. Acad. Sci. Fenn. Ser. A I Math.* **5** (1980), 185–196.

[RI4]   S. RICKMAN: The analogue of Picard's theorem for quasiregular mappings in dimension three, *Acta Math.* **154** (1985), 195–242.

[RI5]   S. RICKMAN: Sets with large local index of quasiregular mappings in dimension three, *Ann. Acad. Sci. Fenn. Ser. A I Math.* **10** (1985), 493–498.

[RI6]   S. RICKMAN: Quasiregular Mappings, *(to appear)*.

[RI7]   S. RICKMAN: Nonremovable Cantor sets for bounded quasiregular mappings, *Preprint 42, Institut Mittag-Leffler, 1989/90*.

[RS]   S. RICKMAN and U. SREBRO: Remarks on the local index of quasiregular mappings, *J. Analyse Math.* **46** (1986), 246–250.

[SA]   J. SARVAS: Coefficients of injectivity for quasiregular mappings, *Duke Math. J.* **43** (1976), 147–158.

[ST]   S. STOÏLOW: Leçons sur les Principles Topologique de la Theorie des Functions Analytiques, Gauthier–Villars, 1938.

[V1]   J. VÄISÄLÄ: Discrete open mappings on manifolds, *Ann. Acad. Sci. Fenn. Ser. A I* **392** (1966), 1–10.

[V2]   J. VÄISÄLÄ: Lectures on $n$-Dimensional Quasiconformal Mappings, Lecture Notes in Math. Vol. **229**, Springer-Verlag, Berlin–Heidelberg–New York, 1971.

[V3]   J. VÄISÄLÄ: A survey of quasiregular maps in $R^n$, Proc. Internat. Congr. Math. (Helsinki, 1978), Vol. **2**, 685–691, Acad. Sci. Fennica, Helsinki, 1980.

[VU1]   M. VUORINEN: Lindelöf–type theorems for quasiconformal and quasiregular mappings, Proc. Complex Analysis Semester, Banach Center Public., Vol. **11**, 353–362, Warsaw, 1983.

[VU2]   M. VUORINEN: Conformal Geometry and Quasiregular Mappings, Lecture Notes in Math. Vol. **1319**, Springer–Verlag, Berlin–Heidelberg–New York, 1988.

[VU3]   M. VUORINEN: On Picard's theorem for entire quasiregular mappings, *Proc. Amer. Math. Soc.* **15** (1989), 383–394.

[ZO1]   V. A. ZORICH: The theorem of M. A. Lavrent'ev on quasiconformal mappings in space (Russian), *Mat. Sb.* **74** (1967), 417–433.

[ZO2]   V. A. ZORICH: Isolated singularities of mappings with bounded distortion (Russian), *Mat. Sb.* **81** (1970), 634–638.

Quasiconformal Space Mappings
– A collection of surveys 1960-1990
Springer–Verlag (1992), 119–131
Lecture Notes in Mathematics Vol. 1508

# DOMAINS AND MAPS

Jussi Väisälä

University of Helsinki, SF–00100 Helsinki, Finland

## 1. Introduction

First I wish to thank Olli Martio for his kind permission to use the title of this survey. Originally, it was the title of a colloquium arranged by him in Jyväskylä in 1989.

The concepts considered in this article are more or less related to quasiconformality. We shall consider the following domains:

uniform, John, QED, LLC, quasiconvex, broad, quasiballs, quasiconformal balls, and the following maps:

quasiconformal, quasisymmetric, weakly quasisymmetric, relatively quasisymmetric, quasimöbius, relatively quasimöbius, quasi–isometric, bilipschitz, coarsely bilipschitz, solid, quasihyperbolic, coarsely quasihyperbolic, freely quasiconformal, pseudoisometric.

My purpose is to give the definitions and the most important results concerning the relations among these properties.

## 2. Domains

2.1. *Notation.* Throughout the paper, $X$ and $Y$ are metric spaces with distance written as $|a - b|$. The one-point extension of $X$ is the Hausdorff space $\dot{X} = X \cup \{\infty\}$, where the neighborhoods of $\infty$ are the complements of bounded sets. The euclidean $n$–space is $R^n$, and $G$ and $G'$ will be domains in $\dot{R}^n$. We always assume that $n \geq 2$.

Open balls are written as $B(x,r)$ and closed balls as $\overline{B}(x,r)$. For real numbers $a$ and $b$, we set

$$a \wedge b = \min(a,b), \quad a \vee b = \max(a,b).$$

The parameter $c$ appearing in various definitions is always assumed to satisfy $c \geq 1$.

**2.2. Quantitativeness.** Suppose that $A$ and $A'$ are conditions containing data $v$ and $v'$, respectively. We say that $A$ implies $A'$ *quantitatively* if $A$ implies $A'$ with $v'$ depending only on $v$. A letter appearing in both $v$ and $v'$ need not have the same value in both conditions. If the condition $A$ deals with a quantity in $R^n$, the dimension $n$ belongs to the data of $A$.

For example, if $f : G \to G'$ is a $K$–quasiconformal map of a $c$–uniform domain onto a $c$–LLC domain, then Theorem 3.8 implies that $G'$ is $c_1$–uniform with $c_1 = c_1(c,K,n)$.

**2.3. Uniformity conditions.** Let $\gamma \subset R^n$ be a rectifiable arc with endpoints $a_1, a_2$. Let $x$ be a point of $\gamma$ dividing $\gamma$ into subarcs $\gamma_1, \gamma_2$. We write

$$\rho(x,\gamma) = l(\gamma_1) \wedge l(\gamma_2),$$

where $l$ is the length. The $c$–*cigar* with core $\gamma$ and parameter $c$ is the open set

$$\operatorname{cig}(\gamma, c) = \bigcup \{B(x, \rho(x,\gamma)/c) : x \in \gamma\}.$$

This set is also defined in the case where one endpoint is $\infty$ and $\gamma \setminus \{\infty\}$ is locally rectifiable. Then $\rho(x,\gamma)$ is the length of the subarc of $\gamma$ between $x$ and the finite endpoint.

An arc $\gamma$ satisfies the $c$–*uniformity conditions* in a domain $G$ if

($U_1$)   $\operatorname{cig}(\gamma, c) \subset G$,
($U_2$)   $l(\gamma) \leq c|a_1 - a_2|$.

Alternatively, ($U_1$) can be written as

($U_1'$)   $\rho(x,\gamma) \leq cd(x, \partial G)$

for all $x \in \gamma$.

A domain $G \subset \dot{R}^n$ is $c$–*John*, $c$–*quasiconvex*, or $c$–*uniform* if each pair of points $a_1, a_2 \in G$ can be joined by an arc $\gamma$ satisfying ($U_1$), ($U_2$), or both $c$–uniformity conditions, respectively. A domain is uniform if it is $c$–uniform for some $c \geq 1$, and the quasiconvex and John domains are defined analogously.

A $c$–uniform domain is trivially $c$–John and $c$–quasiconvex. Bounded convex domains are always uniform. A parallel strip in $R^2$ is convex but not John. A quasiconvex John domain which is not uniform can be obtained by deleting a suitable countable set from a radius of a disk. A simply connected proper subdomain of $R^2$ is uniform if and only if it is a quasidisk [MS, 2.24].

**2.4. Remarks.** There are plenty of alternative characterizations for uniform and John domains. For example, the lengths $l(\gamma_j)$ in the definition of a cigar can be replaced by the diameters $d(\gamma_j)$ or the distances $|x - a_j|$; the length $l(\gamma)$ in ($U_2$) is then replaced by $d(\gamma)$. The article [Vä$_4$] contains a survey of these and some other characterizations. See also 3.19 below and the references [Js$_1$], [Js$_2$], [GO], [Ge$_4$], [GH], [Mn] and [Mo$_1$].

A survey of John domains is given in [NV, Section 2]. One possibility to define a John domain is to use carrots instead of cigars: A domain $G \subset R^2$ is $c$-John if there is $x_0 \in G$, called the *John center*, such that each point $x \in G$ can be joined to $x_0$ by a rectifiable arc $\gamma$ such that for each $z \in \gamma$, the subarc $\gamma_z$ of $\gamma$ between $x$ and $z$ satisfies the condition

$$l(\gamma_z) \le cd(z, \partial G).$$

For bounded domains this is quantitatively equivalent to the definition in 2.3. For other properties of John domains, see [Re$_1$], [Re$_2$, pp. 246 − 253], [Mo$_2$] and [GHM]. Let us only mention that $C^1$-functions in bounded John domains satisfy an inequality of the Poincaré type [Mo$_2$, 3.1]. This leads to *Poincaré* domains; see [Hu] and the references given there.

John domains were first considered by John [Jn, p. 402], uniform domains by Martio–Sarvas [MS] and by Jones [Js$_1$].

In this paper, only finite–dimensional spaces are considered. However, the definitions of 2.3 can be readily extended, for example, to arbitrary Banach spaces. On the other hand, the equivalence of various definitions is then not always valid; see [Vä$_8$, 6.3].

**2.5. QED** *domains.* This concept is based on the modulus of a path family. Let $E$ and $F$ be disjoint continua in a domain $G$. We let $\Delta(E, F; G)$ denote the family of all paths joining $E$ and $F$ in $G$. If

$$M(\Delta(E, F; \dot{R}^n)) \le cM(\Delta(E, F; G))$$

for all pairs $E, F$, the domain $G$ is said to be a *c–quasiextremal distance domain*, briefly a $c$-QED domain.

These domains were introduced by Gehring–Martio [GM$_2$]. They proved that a $c$–uniform domain is $c_1$–QED with $c_1 = c_1(c, n)$. The proof was based on an important result of Jones [Js$_2$] concerning the extension of Sobolev functions. Conversely, a QED domain need not be uniform, as is seen by removing a suitable countable set from a ball. By another result of [GM$_2$], the property $c$-QED quantitatively implies $c$–quasiconvexity.

**2.6.** *Broad domains.* In some sense, broadness is related to QED in the same way as John domains to uniform domains.

The *inner metric* $\delta_G$ of $G$ is defined by

$$\delta_G(a, b) = \inf_{\gamma} d(\gamma)$$

over all arcs $\gamma$ joining $a$ and $b$ in $G$. Let $\varphi : (0, \infty) \to (0, \infty)$ be a decreasing homeomorphism. Let $t > 0$ and let $E, F$ be disjoint continua in $G$ with

$$\delta_G(E, F) \le t(d(E) \wedge d(F)).$$

If

$$M(\Delta(E, F; G)) \ge \varphi(t)$$

for all such pairs $(E, F)$ and for all $t > 0$, we say that $G$ is $\varphi$–*broad*.

From standard modulus estimates it follows that $c$–QED quantitatively implies $\varphi$–broadness. For a simply connected domain in $R^2$, the properties $\varphi$–broad and $c$–John are equivalent [NV, 8.2]. For related results in higher dimensions, see [He].

2.7. *Linear local connectedness.* A set $A$ in a metric space $X$ is *c–linearly locally connected* or *c–LLC* if for each $x \in A$ and each $r > 0$, the following two conditions are satisfied:

(LLC$_1$) Points in $\overline{B}(x, r) \cap A$ can be joined by an arc in $\overline{B}(x, cr) \cap A$.

(LLC$_2$) Points in $A \setminus B(x, r)$ can be joined by an arc in $A \setminus \overline{B}(x, r/c)$.

If $A$ is $c$–LLC, these conditions are in fact valid for all $x \in X$ if $c$ is replaced by $3c$. In this form this concept was introduced by Gehring [Ge$_1$] in 1965. The definition can be extended to sets in $\dot{X}$.

A $c$–QED domain is always $c_1$–LLC with $c_1 = c_1(c, n)$ [GM, 2.7, 2.11]. The domain between two parallel planes in $R^3$ is LLC but not QED.

2.8. *Quasiballs.* A domain $G$ is a *K–quasiball* if $G$ is the image of the unit ball $B^n$ in some $K$–quasiconformal map $f : \dot{R}^n \to \dot{R}^n$. Two–dimensional quasiballs are called *quasidisks,* and they play an important role in several fields. A comprehensive survey of quasidisks is given in [Ge$_3$]. For a simply connected proper subdomain of $R^2$, the properties $K$–quasidisk, $c$–uniform, $c$–QED and $c$–LLC are quantitatively equivalent.

2.9. *Quasiconformal balls.* A domain $G$ is a *K–quasiconformal ball* if there is a $K$–QC map of $G$ onto $B^n$. Trivially, every $K$–quasiball is a $K$–QC ball. For $n = 2$, every simply connected proper subdomain of $R^2$ is a $K$–QC disk with $K = 1$. For $n \geq 3$, no simple characterization for $K$–QC balls is known. A solution for domains of the form $G = D \times R^1 \subset R^3$ is given in [Vä$_6$]. If $n \geq 3$ and if $G$ is a $K$–QC ball, then $\dot{R}^n \setminus G$ is $c$–LLC with $c = c(K, n)$ [Ge$_2$, p. 174]. Hence the domain between two parallel planes in $R^3$ is not a QC ball. On the other hand, the boundary of a QC ball may have positive $n$–measure, even if the boundary is homeomorphic to a sphere [Vä$_5$].

2.10. *Summary.* The following diagram gives the relations between the properties considered above. All implications are quantitative.

$$c\text{-quasiconvex}$$

$$\Uparrow$$

$$K\text{-quasiball} \Rightarrow c\text{-uniform} \Rightarrow c\text{-QED} \Rightarrow c\text{-LLC}$$

$$\Downarrow \qquad\qquad \Downarrow \qquad\qquad \Downarrow$$

$$K\text{-QC ball} \qquad c\text{-John} \qquad \varphi\text{-broad}$$

## 3. Maps

**3.1. *Quasisymmetry and quasimöbius.*** Let $T = (x, a, b)$ be a triple of distinct points in a metric space $X$. The *ratio* of $T$ is the number

$$\rho(T) = \frac{|a - x|}{|b - x|}.$$

Let $\eta : [0, \infty) \to [0, \infty)$ be a homeomorphism. An embedding $f : X \to Y$ is $\eta$–*quasisymmetric* or $\eta$–QS if

(3.2) $$\rho(fT) \leq \eta(\rho(T))$$

for every triple $T$ in $X$.

Analogously, we consider the *cross ratio*

$$\tau(Q) = |a, b, c, d| = \frac{|a - b||c - d|}{|a - c||b - d|}$$

of a quadruple $Q = (a, b, c, d)$ of distinct points in $X$. This is also defined if one of the points is $\infty$. For example,

$$|a, b, c, \infty| = \frac{|a - b|}{|a - c|} = \rho(a, b, c).$$

If $X_0 \subset \dot{X}$, an embedding $f : X_0 \to \dot{Y}$ is $\eta$–*quasimöbius* or $\eta$–QM if

(3.3) $$\tau(fQ) \leq \eta(\tau(Q))$$

for every quadruple $Q$ in $X_0$.

**3.4. *Basic properties.*** The general QS maps were introduced in [TV$_1$] and the QM maps in [Vä$_3$]. See also [As] and [Ri, p. 389].

An $\eta$–QS map is $\theta$–QM with $\theta = \theta_\eta$ [Vä$_3$, 3.2]. If $\infty \in X_0$ and if $f : X_0 \to \dot{Y}$ is $\eta$–QM with $f(\infty) = \infty$, then $f \mid X_0 \setminus \{\infty\}$ is $\eta$–QS. A QM map between bounded spaces is QS, but for a quantitative result the map must be normalized at three points [Vä$_3$, 3.12]. For homeomorphisms between domains in $R^n$, a different normalization is given in [Vä$_3$, 3.24]. Intuitively, a QS map is a QM map fixing the point $\infty$.

If $f : X \to Y$ is $\eta$–QS and if $x, a, b \in X$ with $|a - x| \leq |b - x|$, then

$$|fa - fx| \leq H|fb - fx|$$

with $H = \eta(1)$. An embedding with this property is called *weakly $H$-quasisymmetric*. In some important cases, weak quasisymmetry implies quasisymmetry, for example, if $X$ is a pathwise connected set in $R^n$ and $Y = R^p$ [Vä$_6$, 2.9]. The same is true if both $X$ and $fX$ are quasiconvex [Vä$_7$, 5.5]. A map $f : R^1 \to R^1$ is QS if and only if it satisfies the classical Beurling–Ahlfors condition.

**3.5. *Quasiconformality.*** For the basic theory of *quasiconformal* (QC) maps, we refer to [Vä$_1$]. Suppose that $f : G \to G'$ is a homeomorphism. If $f$ is $\eta$–QM, it follows easily from the metric definition of quasiconformality that $f$ is $K$–QC with $K = \eta(1)^{n-1}$; cf. [Vä$_3$, 5.2]. The converse is not true, since the boundary behavior of a QC map easily destroys the quasimöbius property. One essential difference is that quasiconformality is a *local* property while both quasisymmetry and quasimöbius are *global* properties. In fact, a QC map is locally QS. More precisely, assume that $f : G \to G'$ is a $K$–QC map with $G, G' \subset R^n$ and that $B(x, r) \subset G$. Then for every $t \in (0, 1)$, the restriction $f \mid B(x, tr)$ is $\eta$–QS with $\eta$ depending only on $(K, t, n)$ [Vä$_2$, 2.4]. In fact, it only depends on $(K, t)$ [AVV, 5.23]. More generally, $f$ is $\eta$–QS in every quasihyperbolic ball (see 3.19) of radius $s > 0$ with $\eta$ depending on $(K, s)$ [Vä$_7$, 5.15].

Global results can be obtained if the domains $G$ and $G'$ satisfy certain conditions. The local result above implies that a QC map $f : R^n \to R^n$ is QS. This easily implies that a QC map $f : \dot{R}^n \to \dot{R}^n$ is QM. The following basic result is given in [Vä$_9$], the case $A = G$ in [Vä$_3$, 5.4]; its proof is based on the idea of [GM, 3.1]:

**3.6. Theorem.** *Suppose that $f : G \to G'$ is $K$–QC, that $G'$ is $c$–QED and that $A \subset G$ is $c$–LLC. Then $f \mid A$ is $\eta$–QM and $fA$ is $c_1$–LLC with $(\eta, c_1)$ depending only on $(K, c, n)$.*

*In particular, a QC map between an LLC domain and a QED domain is quasimöbius.*

**3.7. *Invariance theorems.*** The properties uniform and LLC are quasimöbius invariants. More precisely, assume that $f : G \to G'$ is an $\eta$–QM homeomorphism. If $G$ is $c$–uniform, then $G'$ is $c_1$–uniform with $c_1 = c_1(c, \eta, n)$. This follows from [Vä$_3$, 4.9] or from [Vä$_8$, 6.26]. Similarly, if $X_0$ is $c$–LLC and $f : X_0 \to \dot{Y}$ $\eta$–QM, then $fX$ is $c_1$–LLC with $c_1 = c_1(c, \eta)$; see [Vä$_3$, 4.4 and 4.5].

The properties John and broad are not quasimöbius invariants, but they are invariant under quasisymmetric maps.

The following theorem is an easy corollary of 3.6 and the QM invariance of uniform domains:

**3.8. Theorem.** *Suppose that $f : G \to G'$ is $K$–QC and that $G$ is $c$–uniform. Then the following conditions are quantitatively equivalent:*

 (1) *$G'$ is $c$–uniform,*
 (2) *$G'$ is $c$–QED,*
 (3) *$G'$ is $c$–LLC,*
 (4) *$f$ is $\eta$–QM.*

**3.9. *Remarks.*** 1. It follows from 3.8 that a uniform domain cannot be mapped quasiconformally onto an LLC non–uniform domain. In particular, the domains $B^p \times R^{n-p}$ are not QC balls for $1 \le p \le n - 2$.

2. An $\eta$–quasimöbius homeomorphism $f : G \to G'$ always has a homeomorphic extension $\overline{f} : \overline{G} \to \overline{G}'$, which is also $\eta$–QM. Hence a QC map between uniform domains can be extended to a homeomorphism between the closures. A more general result is given in 3.22.4 below.

3. For further related results, see [GM$_1$], [MV], [AT], [Tr], [HK$_{1,2,3}$], [HV].

**3.10. *Subinvariance.*** Let $P(c)$ be a property of domains involving the parameter $c$. We say that $P(c)$ is a QC *subinvariant* if the following statement is true: Suppose that

$f : G \to G'$ is $K$–QC, that $D \subset G$ is a domain and that $D$ and $G'$ have the property $P(c)$. Then $fD$ has $P(c_1)$ with $c_1 = c_1(c, K, n)$.

The first subinvariance result is due to O. Martio, who discovered that the property $c$–QED is a QC subinvariant; cf. [FHM, Theorem 1]. We recall the proof, which is easy but ingenious:

Suppose that $f : G \to G'$ is $K$–QC and that $D \subset G$ and $G'$ are $c$–QED. Let $E'$ and $F'$ be disjoint continua in $fD$, and write $E = f^{-1}E'$, $F = f^{-1}F'$,

$$\Gamma'_D = \Delta(E', F'; fD), \ \Gamma'_G = \Delta(E', F'; G'), \ \Gamma' = \Delta(E', F'; \dot{R}^n),$$

$$\Gamma_D = \Delta(E, F, D), \ \Gamma_G = \Delta(E, F; G), \ \Gamma = \Delta(E, F; \dot{R}^n).$$

Then

$$M(\Gamma') \leq cM(\Gamma'_G) \leq KcM(\Gamma_G) \leq KcM(\Gamma) \leq Kc^2 M(\Gamma_D) \leq K^2 c^2 M(\Gamma'_D).$$

Hence $fD$ is $K^2 c^2$–QED. □

The property of being $c$–uniform is also QC subinvariant. To see this, let $f : G \to G'$ be $K$–QC, and suppose that $D \subset G$ and $G'$ are $c$–uniform. The subinvariance of QED and 2.10 imply that $fD$ is $c_1$–QED. Hence $fD$ is $c_2$–uniform by 3.8 with $c_2 = c_2(c, K, n)$. □

3.11. *Inner quasisymmetry.* Suppose that $f : G \to G'$ is $K$–QC with $G$ $c$–uniform and $G'$ $c$–QED. Then $f$ is $\eta$–QM by 3.8 with $\eta$ depending on $(K, c, n)$. We next give an analogous result where $G$ is assumed to be $c$–John and $G'$ $\varphi$–broad. However, the quasisymmetry of $f$ must be considered in the inner metric $\delta_G$ defined in 2.6. The following result is a special case of [Vä$_6$, 2.20]:

3.12. **Theorem.** *Suppose that $G$ and $G'$ are bounded domains in $R^n$, that $G$ is $c$–John with center $x_0$, that $G'$ is $\varphi$–broad and that $f : G \to G'$ is a $K$–QC map with $d(G') \leq cd(f(x_0), \partial G')$. Then $f$ is $\eta$–QS in the inner metrics $\delta_G$ and $\delta_{G'}$, with $\eta$ depending only on $(c, \varphi, K, n)$.*

3.13. *Relativization.* Sometimes it is natural to consider maps $f : X \to Y$ which do not necessarily satisfy the QS condition for all triples in $X$ but only *relative* to a given subset $A$ of $X$. More precisely, we say that a triple $T = (x, a, b)$ of distinct points in $X$ is a triple in $(X, A)$ if either $x \in A$ or $\{a, b\} \subset A$. An embedding $f : X \to Y$ is said to be $\eta$–QS rel $A$ if (3.2) is valid for every triple $T$ in $(X, A)$.

Analogously, an embedding $f : X_0 \to \dot{Y}$, $X_0 \subset \dot{X}$, is $\eta$–QM rel $A \subset X_0$ if (3.3) is valid for every quadruple $A = (a, b, c, d)$ in $(X_0, A)$, which means that $\{a, d\} \subset A$ or $\{b, c\} \subset A$.

The relative theory is developed in [Vä$_8$, Section 5].

3.14. *Quasi-isometries and related maps.* Let $\varphi : [0, \infty) \to [0, \infty)$ be a homeo-morphism with $\varphi(t) \geq t$. An embedding $f : X \to Y$ is said to be a $\varphi$-*quasi-isometry* if

$$\varphi^{-1}(|x - y|) \leq |fx - fy| \leq \varphi(|x - y|)$$

for all $x, y \in X$. The function $\varphi(t) = Mt$, $M \geq 1$, gives the *M–bilipschitz* maps, which are characterized by the condition

$$|x - y|/M \leq |fx - fy| \leq M|x - y|.$$

If $C \geq 0$ and $f$ satisfies the weaker condition

(3.15) $$(|x - y| - C)/M \leq |fx - fy| \leq M|x - y| + C,$$

we say that $f$ is *C–coarsely M–bilipschitz*.

We analyze the second inequality of (3.15), which is called the *C–coarse M–Lipschitz* condition. It implies that $f$ has the following two properties:

(1) $f$ is Lipschitz for large distances. More precisely, if $|x - y| \geq t_0 > 0$, then $|fx - fy| \leq M_0|x - y|$ where $M_0 = M + C/t_0$.

(2) $f$ quasi preserves reasonable distances. More precisley, $|x - y| \leq t_1$ implies $|fx - fy| \leq M_1$ with $M_1 = Mt_1 + C$.

In certain important cases, each of these conditions is in fact equivalent to the coarse Lipschitz condition:

**3.16. Lemma.** *Suppose that $f : X \to Y$ is a map and that one of the following conditions is true:*

(1) *$X$ is connected, $d(X) = \infty$, $t_0 \geq 0$, $M_0 \geq 0$, and*

$$|fx - fy| \leq M_0|x - y|$$

*whenever $|x - y| \geq t_0$.*

(2) *$X$ is $c$–quasiconvex, $t_1 > 0$, $M_1 \geq 0$, and*

$$|fx - fy| \leq M_1$$

*whenever $|x - y| \leq t_1$.*

*Then*

(3.17) $$|fx - fy| \leq M|x - y| + C$$

*for all $x, y \in X$ with $(M, C)$ depending only on the given data.*

*Proof.* Assume that (1) is true and that $x, y \in X$. If $|x - y| \leq t_0$, then (3.17) holds with $M = M_0$, $C = 0$. Suppose that $|x - y| \geq t_0$. Choose a point $z \in X$ with $|z - x| = 2t_0$. Then $t_0 \leq |y - z| \leq 3t_0$, and we obtain

$$|fx - fy| \leq |fx - fz| + |fz - fy| \leq M_0|x - z| + M_0|z - y| \leq 5M_0t_0,$$

and (3.17) holds with $M = 0$, $C = 5M_0t_0$.

Next assume that (2) is true and that $x, y \in X$. Join $x$ and $y$ by an arc $\gamma$ with $l(\gamma) \leq c|x - y|$. Let $k \geq 0$ be the unique integer with $kt_1 \leq l(\gamma) < (k + 1)t_1$. Choose successive points $x = x_0, \ldots, x_{k+1} = y$ of $\gamma$ such that the subarc of $\gamma$ between each pair

$x_{j-1}, x_j$ has length at most $t_1$. Then $|x_{j-1} - x_j| \leq t_1$, and hence $|fx_{j-1} - fx_j| \leq M_1$. This implies that

$$|fx - fy| \leq (k+1)M_1 \leq M_1 l(\gamma)/t_1 + M_1 \leq cM_1 |x - y|/t_1 + M_1,$$

which gives (3.17) with $M = cM_1/t_1$, $C = M_1$. ⊡

3.18. *Remarks.* 1. It follows from 3.16 that a bijective quasi–isometry between quasiconvex spaces is always coarsely bilipschitz.
2. An $M$–bilipschitz map is always $\eta$–QS with $\eta(t) = M^2 t$. An $M$–bilipschitz homeomorphism $f : G \to G'$ is always $K$–QC with $K = M^{2n-2}$, but the bilipschitz maps form a very special subclass of the QC maps. The situation is essentially different if we consider the quasihyperbolic metric instead of the euclidean metric; see 3.22.2.

3.19. *Quasihyperbolic metric.* The quasihyperbolic or QH metric was introduced by F.W. Gehring in the seventies as a natural generalization of the classical hyperbolic metric of a half space. Let $G$ be a proper subdomain of $R^n$. For $x \in G$ we write $\delta(x) = d(x, \partial G)$. The QH length of a rectifiable arc $\gamma \subset G$ is the line integral

$$l_k(\gamma) = \int_\gamma \frac{|dx|}{\delta(x)},$$

and the QH *distance* between points $a, b \in G$ is the number

$$k_G(a, b) = \inf_\gamma l_k(\gamma)$$

over all $\gamma$ joining $a$ and $b$ in $G$. This metric has turned out to be a useful tool in many questions. For example, uniform domains can be characterized by the condition

$$k_G(a, b) \leq c \log \left(1 + \frac{|a - b|}{\delta(a) \wedge \delta(b)}\right);$$

see [GO], [Vu$_1$, p. 85] and [Ge$_4$, p. 99].

3.20. *Three classes of maps.* Let $G$ and $G'$ be proper subdomains of $R^n$ and let $f : G \to G'$ be a homeomorphism. We consider $f$ as a map between the metric spaces $(G, k_G)$ and $(G', k_{G'})$. If this map is a $\varphi$–quasi-isometry, $M$–bilipschitz, or $C$–coarsely $M$–bilipschitz, we say that $f$ is $\varphi$–*solid*, $M$–*quasihyperbolic* ($M$–QH) or $C$–*coarsely $M$–quasihyperbolic* (($M, C$)–CQH), respectively.
Relations between these concepts and QC maps are given in the following result:

3.21. **Theorem.** *Suppose that $G, G'$ are proper subdomains of $R^n$. Then the following quantitative implications are true for a homeomorphism $f : G \to G'$.*

$$M\text{-QH} \Rightarrow K\text{-QC} \Rightarrow \varphi\text{-solid} \Rightarrow (M, C)\text{-CQH}.$$

*Proof.* The first implication follows easily with $K = M^{2n-2}$ from the fact that the expression $k(x, y)\delta(x)/|x-y|$ is close to 1 if $y$ is in a small neighborhood of $x$. The second one was proved by Gehring and Osgood [GO, p. 62]. One can choose $\varphi(t) = a_K(t \vee t^{1/K})$; see [Vu$_2$, 12.20]. Since the metric space $(G, k_G)$ is 1-quasiconvex, the last implication follows from the part (2) of 3.16. One can choose $M = C = \varphi(1)$. □

**3.22. Remarks.** 1. The converse implications of 3.21 are false. However, if $f : G \to G'$ is a homeomorphism such that the restriction $f_D : D \to fD$ of $f$ to each subdomain $D \subset G$ is $\varphi$-solid, then $f$ is $K$–QC with $K = K(\varphi, n)$ [TV$_2$, 6.12]. Maps with this property are called *freely $\varphi$-quasiconformal*, and they give the starting point for the theory of quasiconformality in arbitrary Banach spaces [Vä$_{7,8,10}$]. It is also true that if each $f_D$ is $(M, C)$–CQH, then $f$ is $K$–QC with $= K(M, C, n)$ [Vä$_{10}$]. An early version of these results was given by Gehring [Ge$_1$, p. 14] in terms of $\theta$–maps.

2. The QH maps form an important subclass of the QC maps. If $n \neq 4$, the QC maps and, more generally, the solid maps can be quantitatively approximated by the QH maps [TV$_2$, 7.4]. The Beurling–Ahlfors extension of a QS map of $R^1$ is QH in the half plane.

3. A map $f : G \to G'$ satisfying the condition

$$(k_G(x, y) - C)/M \le k_{G'}(fx, fy) \le M k_G(x, y)$$

is called an $(M, c)$-*pseudoisometry*. These maps appear in Mostow's ridigity theory [Mw, p. 66], [Th, p. 5.39], and they form a subclass of the CQH maps.

4. Suppose that $G$ and $G'$ are uniform proper subdomains of $R^n$. Recall from 3.8 that a QC map $f : G \to G'$ is always QM. A CQH map $f : G \to G'$ need not be QM, but it extends to a homeomorphism $\overline{f} : \overline{G} \to \overline{G}'$ which is QM rel $\partial G$ [Vä$_8$, 7.9]. In particular, $\overline{f} \mid \partial G$ is QM. A CQH map $f : B^n \to B^n$ with $f(0) = 0$ extends to $\overline{f} : \overline{B}^n \to \overline{B}^n$, and $\overline{G}$ is QS rel $\partial B^n$. These results are quantitative and extend to arbitrary Banach spaces.

5. Since a QM homeomorphism $f : G \to G'$ is always QC, it is natural to ask whether a homeomorphism $\overline{f} : \overline{G} \to \overline{G}'$ which is QM rel $\partial G$ is CQH in $G$. The answer is affirmative, but the CQH parameters $M, C$ depend heavily on $n$, and the free version in Banach spaces is only true for a special class of domains which includes all uniform domains [Vä$_{10}$].

6. One can show that a homeomorphism $f : G \to G'$ is $\eta$–QS in the QH metric if and only if, quantitatively, $f$ is $K$–QC; see [Vä$_7$, 3.5].

# References

[AVV] G.D. Anderson, M.K. Vamanamurthy and M. Vuorinen, Dimension–free quasiconformal distortion in $n$–space, Trans. Amer. Math. Soc. 297, 1986, 687–706.

[As] V.V. Aseev, Quasisymmetric embeddings and maps of bounded modulus distortion, Viniti 7190–84, Moscow, 1984. (Russian)

[AT] V.V. Aseev and D.A. Trocenko, Quasisymmetric embeddings, quadruples of points and modulus distortion, Sibirsk. Mat. Zh. 28, 1987, 32–38. (Russian)

[FHM] J.L. Fernández, J. Heinonen and O. Martio, Quasilines and conformal mappings, J. Analyse Math. 52, 1989, 117–132.

[Ge₁] F.W. Gehring, The Carathéodory convergence theorem for quasiconformal mappings, Ann. Acad. Sci. Fenn. Ser. A I Math. 336/11, 1963, 1–21.

[Ge₂] — Extension of quasiconformal mappings in three space, J. Analyse Math. 14, 1965, 171–182.

[Ge₃] — Characteristic properties of quasidisks, Séminaire de Mathématiques Supérieures, Montréal, 1982.

[Ge₄] — Uniform domains and the ubiquitous quasidisk, Jahresber. Deutsch. Math.-Verein. 89, 1987, 88–103.

[GH] F.W. Gehring and K. Hag, Remarks on uniform and quasiconformal extension domains, Complex Variables 9, 1987, 175–188.

[GHM] F.W. Gehring, K. Hag and O. Martio, Quasihyperbolic geodesics in John domains, Math. Scand. 65, 1989, 75–92.

[GM₁] F.W. Gehring and O. Martio, Lipschitz classes and quasiconformal mappings, Ann. Acad. Sci. Fenn. Ser. A I Math. 10, 1985, 203–219.

[GM₂] — Quasiextremal distance domains and extension of quasiconformal mappings, J. Analyse Math. 45, 1985, 181–206.

[GO] F.W. Gehring and B.G. Osgood, Uniform domains and the quasihyperbolic metric, J. Analyse Math. 36, 1979, 50–74.

[He] J. Heinonen, Quasiconformal mappings onto John domains, Rev. Mat. Iberoamericana 5, 1989, 97–123.

[HK₁] D. Herron and P. Koskela, Quasiextremal distance domains and extendability of quasiconformal mappings, Complex Variables 15, 1990, 167–179.

[HK₂] — Uniform, Sobolev extension and quasiconformal circle domains, J. Analyse Math., to appear.

[HK₃] — Uniform and Sobolev extension domains, Proc. Amer. Math. Soc., to appear.

[HV] D. Herron and M. Vuorinen, Positive harmonic functions in uniform and admissible domains, Analysis 8, 1988, 187–206.

[Hu] R. Hurri, Poincaré domains in $R^n$, Ann. Acad. Sci. Fenn. Ser. A I Math. Diss. 71, 1988, 1–42.

[Jn] F. John, Rotation and strain, Comm. Pure Appl. Math. 14, 1961, 319–413.

[Js₁] P.W. Jones, Extension theorems for BMO, Indiana Univ. Math. J. 29, 1980, 41–66.

[Js₂] — Quasiconformal mappings and extendability of Sobolev spaces, Acta Math. 147, 1981, 71–88.

[Mn]    G. Martin, Quasiconformal and bi-Lipschitz homeomorphisms, uniform domains and the quasihyperbolic metric, Trans. Amer. Math. Soc. 292, 1985, 169–191.

[Mo$_1$]    O. Martio, Definitions for uniform domains, Ann. Acad. Sci. Fenn. Ser. A I Math. 5, 1980, 197–205.

[Mo$_2$]    — John domains, bilipschitz balls and Poincaré inequality, Rev. Roumaine Math. Pures Appl. 33, 1988, 107–112.

[MS]    O. Martio and J. Sarvas, Injectivity theorems in plane and space, Ann. Acad. Sci. Fenn. Ser. A I Math. 4, 1979, 383–401.

[MV]    O. Martio and M. Vuorinen, Whitney cubes, $p$-capacity and Minkowski content, Exposition. Math. 5, 1987, 17–40.

[Mw]    G.D. Mostow, Strong rigidity of locally symmetric spaces, Ann. Acad. Studies 78, Princeton University Press, 1973.

[NV]    R. Näkki and J. Väisälä, John disks, Exposition. Math. 9, 1991, 3–44.

[Re$_1$]    Y.G. Reshetnyak, Stability in Liouville's theorem on conformal mappings of a space for domains with nonsmooth boundary, Sibirsk. Mat. Zh. 17, 1976, 361–369. (Russian)

[Re$_2$]    — Space mappings with bounded distortion, Amer. Math. Soc., 1989.

[Ri]    S. Rickman, Quasiconformally equivalent curves, Duke Math. J. 36, 1969, 387–400.

[Th]    W.P. Thurston, The geometry and topology of three-manifolds, mimeographed notes, Princeton University, 1980.

[Tr]    D.A. Trocenko, Quasiconformal and quasi-isometric reflections, Dokl. Akad. Nauk SSSR 287, 1986, 1067–1071. (Russian)

[TV$_1$]    P. Tukia and J. Väisälä, Quasisymmetric embeddings of metric spaces, Ann. Acad. Sci. Fenn. Ser. A I Math. 5, 1980, 97–114.

[TV$_2$]    — Lipschitz and quasiconformal approximation and extension, Ibid. 6, 1981, 303–342.

[Vä$_1$]    J. Väisälä, Lectures on $n$-dimensional quasiconformal mappings, Lecture Notes in Mathematics 229, Springer–Verlag, 1971.

[Vä$_2$]    — Quasisymmetric embeddings in euclidean spaces, Trans. Amer. Math. Soc. 264, 1981, 191–204.

[Vä$_3$]    — Quasimöbius maps, J. Analyse Math. 44, 1984/85, 218–234.

[Vä$_4$]    — Uniform domains, Tôhoku Math. J. 40, 1988, 101–118.

[Vä$_5$]    — Quasiconformal maps and positive boundary measure, Analysis 9, 1989, 205–216.

[Vä$_6$]    — Quasiconformal maps of cylindrical domains, Acta Math. 162, 1989, 201–205.

[Vä$_7$]    — Free quasiconformality in Banach spaces I, Ann. Acad. Sci. Fenn. Ser. A I Math. 15 1990, 355–379.

[Vä$_8$]    — Free quasiconformality in Banach spaces II, preprint.

[Vä$_9$]    — Bounded turning and quasiconformal maps, preprint.

[Vä$_{10}$]    — Free quasiconformality in Banach spaces III, in preparation.

[Vu$_1$]    M. Vuorinen, Conformal invariants and quasiregular mappings, J. Analyse Math. 45, 1985, 69–114.

[Vu$_2$]   —   Conformal geometry and quasiregular mappings, Lecture Notes in Math. 1319, Springer–Verlag, 1988.

Quasiconformal Space Mappings
– A collection of surveys 1960–1990
Springer–Verlag (1992), 132–148
Lecture Notes in Mathematics Vol. 1508

# THE GLOBAL HOMEOMORPHISM THEOREM
# FOR SPACE QUASICONFORMAL MAPPINGS,
# ITS DEVELOPMENT
# AND RELATED OPEN PROBLEMS

## V. A. Zorich

Moscow University, 119899 Moscow, Russia

Dedicated to Gerhard and Inge Krautschneider

### Abstract

The global homeomorphism theorem is the following specifically multidimensional phenomenon: any locally homeomorphic quasiconformal mapping $f : R^n \to R^n$ is a bijection if $n \geq 3$.

We present here a brief review of results and open problems related to this theorem.

# 1   Conformal and quasiconformal mappings

The linear mapping $L : R^n \to R^n$ transforms the unit ball $B^n$ into the ellipsoid $L(B^n)$. The value of the ratio $k_L$ of the major semiaxis to the minor semiaxis (or the value $\log k_L$) may be considered as a measure of the nonconformality of $L$. When $k_L = 1$ ($\log k_L = 0$) the mapping $L$ is conformal.

If $f : D \to R^n$ is a diffeomorphism of a domain $D \subset R^n$, then the value $k_{f'(x)}$ measures the deviation of $f$ from conformality at a point $x \in D$, and $k_f = \sup_{x \in D} k_{f'(x)}$ can be regarded as a measure of the nonconformality of $f$ in the domain $D$.

1980 *Mathematics Subject Classification* (1985 *Revision*). Primary 30C60, secondary 53A30.
*Keywords and phrases.*Quasiconformal mapping, Riemannian manifold, nonlinear operator, injectivity, invertibility, global homeomorphism, conformal capacity, modulus.

We now recall that a mapping $f : D \rightarrow R^n$ of a domain $D \subset R^n$ is called $k$-*quasiconformal* in $D$ if the quantity

$$k_f(x) := \limsup_{\delta \rightarrow 0+} \frac{\max\{|f(\xi) - f(x)| : |\xi - x| = \delta\}}{\min\{|f(\xi) - f(x)| : |\xi - x| = \delta\}}$$

(called the *linear dilatation* or *coefficient of quasiconformality* of $f$ at $x \in D$) is bounded in $D$ and $k_f(x) \leq k$ for almost all points $x \in D$. Note that $k = 1$ if and only if $f$ is a conformal mapping. A mapping is called *quasiconformal* or a *mapping with bounded dilatation* if it is $k$-quasiconformal for some $k$ $(1 \leq k < \infty)$. Here we shall mainly use this definition for homeomorphic or for locally homeomorphic mappings. The definition of quasiconformality has a natural generalization to mappings of metric spaces and to mappings of Riemannian manifolds; we shall need this more general notion later. The coefficient of quasiconformality is obviously invariant under a conformal change of metric.

Quasiconformal mappings became an object of mathematical investigation and application in the 20th to 30th years of our century in the papers of Grötzsch [Gr1], [Gr2], Lavrent'ev [Lav1], Ahlfors [A1], and Teichmüller [T1], [T2]. The term "quasiconformal mapping" is due to Ahlfors; he introduced it in his famous investigation of the geometric aspects of Nevanlinna theory.

## 2 The role of dimension

Quasiconformal mappings in dimensions $n = 2$ and $n \geq 3$ have many common functional and geometric properties.

For example, in both cases the following frequently used theorem [Lav1], [B2], [A2], [G2], [V3], [Vu2], [AVV2], [Vu3] on boundedness of distortion is valid: Distortion under quasiconformal mappings is bounded not only locally but also globally. Here follows a concrete version of this property:

Let $Q_k(R^n)$ be the class of one-to-one $k$-quasiconformal mappings $f : R^n \rightarrow R^n$ of the whole space $R^n$ $(n \geq 2)$. Let

$$M_f(r) := \sup_{|x|=r} |f(x) - f(0)|, \qquad m_f(r) := \inf_{|x|=r} |f(x) - f(0)|.$$

**Theorem** (on the coefficient of distortion) *The function*

$$\epsilon(n, k) := \sup_{f \in Q_k(R^n)} \sup_{r > 0} \frac{M_f(r)}{m_f(r)} \tag{1}$$

*is finite for all values of the arguments $(n \in N, n \geq 2$, and $k \in R, k \geq 1)$.*

But at the same time there is a great difference in the properties of conformal and quasiconformal mappings in dimensions $n = 2$ and $n \geq 3$. The nature of the difference was discovered essentially in the middle of the 19th century, when almost simultaneously Riemann stated his existence theorem on the richness of conformal mappings in dimension $n = 2$ and when four years earlier Liouville discovered the conformal rigidity of domains in $R^n$ when $n \geq 3$. More precisely, Liouville proved that the following theorem is valid.

**Theorem** (on conformal mappings in space) *Any conformal mapping ($C^4$-smooth) of a domain in the space $R^n$ of dimension $n > 2$ is an element of the Möbius group.*

(Recall that the Möbius group consists of compositions of translations, dilatations, and inversions. This group can be generated by reflections in spheres or planes.)

The reason for this phenomenon is that when $n > 2$ the condition $f'(x) \in \Lambda O(n)$ (meaning that the matrix $f'(x)$ is a scalar multiple of an orthogonal matrix) generates an overdetermined system of $(n(n+1)/2) - 1$ equations in $n$ functions. When $n = 2$ the system is well defined: this is the classical Cauchy–Riemann system.

One can find various proofs of the Liouville theorem, e.g. in [G2], [Re2], [Re3], [BI], [NR2].

Under some minimal regularity conditions and even without any a priori injectivity condition of the mapping $f$ the same theorem was proved by Yu. G. Reshetnyak [Re1]. See also new results of Iwaniec and Martin [IM].

A large number of investigations [Lav3], [Lav4], [Lav5], [B2], [B3], [Re2], [Re3], [K], [Se], [I1] (it might have been an object of special survey) were carried out in connection with the goal [Lav3] of studying the stability problem in Liouville's theorem. For later use and to clarify this problem, we now present here just one such result [B1], though not in the most general formulation.

Let $Q_{1+\epsilon}$ be the class of quasiconformal mappings $f : B^n \to R^n$ of the ball $B^n, n \geq 2$, such that $k_f \leq 1 + \epsilon$. Let $Q_1$ denote the class of conformal mappings of $B^n$.

**Theorem** (on stability of conformal mappings) *For any fixed $n \geq 2$ the function*

$$\mu(n, \epsilon) := \sup_{f \in Q_{1+\epsilon}} \; \inf_{\varphi \in Q_1} \; \sup_{x \in B^n} |\varphi^{-1} \circ f(x) - x|$$

*tends to zero as $\epsilon \to 0+$.*

Up to now the stability of conformal mappings has been studied in various domains and in various norms [Re2], [Re3]. One has also an asymptotic estimate

$$\mu(n, \epsilon) \leq \nu(n)\epsilon \tag{2}$$

as $\epsilon \to 0+$, with coefficient $\nu(n)$ depending on $n$.

Thus, in a sense, the stability of conformal mappings takes place in any space $R^n$, $n \geq 2$. But the corollaries of the stability theorem in dimensions $n = 2$ and $n > 2$ are different because of the difference in these dimensions between the classes of conformal mappings as stated in the theorems of Riemann and Liouville.

Finally we point out that according to R. Nevanlinna [NR2] the Liouville theorem is also valid for conformal mappings of Hilbert spaces. Still any publications on infinite-dimensional stability in Liouville's theorem are not known to me.

After Liouville's discovery described above the first paper concerning specifically multi-dimensional effects of quasiconformality was the paper [Lav2] of M. A. Lavrent'ev. It contained two assertions, which can be regarded now as being proved after some corrections.

One of Lavrent'ev's assertions is the following.

*Let $f : B^3 \to B^3$ be a quasiconformal mapping of the open unit ball into itself. If the closure of the image $f(B^3)$ contains the boundary sphere $\partial B^3$, then $f(B^3) = B^3$.*

Strictly speaking, this assertion is not true. But after investigations on the boundary behavior of quasiconformal mappings [V1], [Zo1], [GV] (see also [N1], [N2], [Vu1]) we know that if, e.g., in addition $f(B^3)$ is locally connected at every point $x \in \partial B^3$, then really $B^3 \setminus f(B^3)$ is empty.

In the two-dimensional case the additional local connectivity hypothesis stated above makes the assertion trivially valid for any homeomorphism $f$, even without assumptions of its quasiconformality. But in dimension 3 this assertion is really meaningful and nontrivial. In particular, it follows that the ball $B^3$ and the ball with a radius removed are not even quasiconformally equivalent. This is in contrast to the possibility of performing even conformal mappings onto canonical domains in the two-dimensional case. Thus we see: though quasiconformal mappings constitute a much wider class than the class of conformal mappings in space (e.g. any diffeomorphism restricted to a compact subset of its domain of definition is quasiconformal), they still retain some properties connected with conformal rigidity of space domains.

But the the main assertion in [Lav2] was as follows.

*If $f : R^3 \to R^3$ is a locally homeomorphic quasiconformal mapping of 3-dimensional Euclidean space $R^3$ into itself, then $f$ is a homeomorphism and, moreover, a homeomorphism onto the whole space $R^3$.*

This statement turned out to be quite correct [Zo2]. We are now ready to proceed to the main topic of the present survey.

# 3 Global homeomorphism theorem

The following theorem is valid [Zo2], [Zo3]:

**Theorem (on global homeomorphism)** *Let $f : R^n \to R^n$ be a locally homeomorphic mapping with bounded dilatation. If $n > 2$, then the mapping $f$ is a globally homeomorphic quasiconformal bijection.*

Clearly this theorem may be reformulated in terms of the nonlinear equation $f(x) = y$: *under the indicated conditions on $f : R^n \to R^n$, for every value of the right side $y \in R^n$ there exists a unique $x \in R^n$ such that $f(x) = y$.*

The condition $n > 2$ is essential, as the mapping $z \mapsto \exp(z)$ illustrates.

It is also essential that $f$ be a local homeomorphism, as shown by the example of the mapping $f : R^3 \to R^3 \setminus \{0\}$ with bounded dilatation constructed in [Zo3]. This doubly periodic mapping is similar to $\exp(z)$, but in contrast to $\exp(z)$ it has branchings on a doubly periodic system of parallel lines. The mapping constructed has two omitted values (0 and $\infty$), just as $\exp(z)$ has. In this connection, a natural question [Zo3] arises — a question on Picard-type theorems for entire (defined in the whole space $R^n$) mappings with bounded dilatation. An unexpected answer to this question was obtained by Rickman, who proved that the set of omitted points is always finite [Ri1], but in contrast to the two-dimensional case the number of omitted points depends on the coefficient $k_f$ of quasiconformality of the mapping $f$ and can increase without bound as $k_f$ increases [Ri2].

A specifically multi-dimensional ($n \geq 3$) effect also takes place when $k_f \searrow 1$. It turns out [Go], [MRV], [I1] that if $k_f$ is sufficiently close to 1, then the space mapping $f$ has no branch points, in contrast to the two-dimensional case. Because of the theorem on global homeomorphism this means that the mapping $f : R^n \to R^n$ with coefficient of quasiconformality close to 1 is a homeomorphism and moreover a homeomorphism "onto" ($f(R^n) = R^n$), i.e., $f$ has no finite omitted values at all.

The first of the two statements of the global homeomorphism theorem, that the mapping is injective in the large, is the most important one. The other statement, on surjectivity, is easily proved using the modulus method (method of extremal lengths or conformal capacities) in both the two-dimensional and space cases, and hence does not reflect specific space properties.

# 4 Modulus (extremal length) method and some remarks on the proof of the theorem

The method of (quasi)conformal (quasi)invariants, or more precisely the modulus (extremal lengths or conformal capacities) method is an important tool for study of quasiconformal mappings. Its roots come from the geometric theory of functions. In explicit conformally invariant form this method was formulated by Ahlfors and Beurling [AB], and then it was extended to higher dimensions by Fuglede [F]. After the article [Lo] (containing the concept of conformal capacity), and the papers [Sh], [G1], [G2], [V1], [V2], [Zo1] (developing the equivalent concept of extremal length and modulus), it became one of the main tools for studying space quasiconformal mappings (see [V3], [Ca], [Sy], [Vu2]).

Let $\Gamma$ be a family of curves in $R^n$ or in any other Riemannian manifold $X$, and let $A(\Gamma)$ be a family of Borel-measurable nonnegative functions $\rho$ on $X$ such that $\int_\gamma \rho \geq 1$ for any locally rectifiable curve $\gamma \in \Gamma$. Then the quantities

$$M(\Gamma) = \inf_{\rho \in A(\Gamma)} \int_X \rho^n \quad \text{and} \quad \lambda(\Gamma) = M(\Gamma)^{1/(1-n)}$$

are called the *modulus* and *extremal length*, respectively, of the family $\Gamma$.

It is clear that the modulus is an invariant of the class of conformal metrics on $X$. In fact, multiplication by $\alpha(x)$ of the length element at a point $x \in X$ multiplies the volume element by $\alpha^n(x)$. Instead of the function $\rho$ we can now take a new admissible function $\rho/\alpha$. An analogous construction shows that quasiconformal change of the metric cannot cause a change of $M(\Gamma)$ by more than a factor $k^{n-1}$. In particular, the following relations hold between the modulus $M(\Gamma)$ of any family of curves $\Gamma \subset X$ and the modulus of its image $f(\Gamma)$ under a $k$-quasiconformal mapping $f : X \to Y$ of an $n$-dimensional Riemannian manifold:

$$k^{1-n} M(\Gamma) \leq M(f(\Gamma)) \leq k^{n-1} M(\Gamma). \tag{$*$}$$

So, the modulus is (quasi)conformal (quasi)invariant.

In the theory of quasiconformal mappings one also uses a notion of conformal capacity or more often a notion of conformal capacity of a condenser. By a condenser in $X$ one means any pair of disjoint subsets $E_0, E_1$ in $X$. Usually $E_0, E_1$ are continua.

The conformal capacity of a condenser is defined as

$$cap(E_0, E_1) = \inf_{U} \int_X \| \text{ grad } U \|^n,$$

where the infimum is taken over all nonnegative functions in the class $W_n^1$, equal to 0 on $E_0$ and equal to 1 on $E_1$. Here $n = \dim X$.

If $E_0$ and $E_1$ are continua then the quantity $cap(E_0, E_1)$ coincides with the modulus $M(\Gamma)$ of the family of curves $\Gamma$ connecting the sets $E_0$ and $E_1$ in $X$ [Zi], [He].

Let $X$ be a complete noncompact Riemannian manifold, e.g., $R^n$ or the Lobachevski space $H^n$. We fix a bounded neighborhood of some point and consider the family $\Gamma_\infty$ of all curves in $X$ passing from this neighborhood to infinity. It does not depend on the choice of the original neighborhood whether $M(\Gamma_\infty) = 0$ or $M(\Gamma_\infty) \neq 0$ . Hence, $M(\Gamma_\infty)$ is a discrete conformal invariant of the manifold $X$, which it is natural to call the *conformal capacity of $X$ at infinity*. Thus the capacity of $R^n$ at infinity is equal to zero ($\partial R^n$ is a point in the stereographic model of $R^n$). The capacity of $H^n$ at infinity is positive (in the conformal model of Poincaré the Lobachevski space $H^n$ is a ball $B^n \subset R^n$ and $\partial H^n = \partial B^n = S^{n-1}$ is a sphere).

To solve qualitative problems one usually uses the quasi-invariance of the modulus (∗) also only qualitatively. Analyzing the proof of the global homeomorphism theorem, one can note that besides the topological conditions the only condition the manifold and the mapping should obey is as follows: if $M(\Gamma) = 0$ then $M(f(\Gamma)) = 0$. It is even sufficient that $M(f(\Gamma)) = 0$ for families $\Gamma$ contained in the domains of one-to-one-ness of $f$ and going to infinity in $R^n$. This fact leads to one of the generalizations of the theorem, which is discussed in the next section.

Proofs of injectivity theorems usually have a starting topological part and an ending metric part. In the topological part one usually constructs (cf. [Hd], [C], [NF], [NR1], [Zo3], [J2], [J3], [MRV]) a covering over a curve, over a surface, or over a domain, which is then to be extended continuously in the space of the range. If, e.g., a mapping $f : R^n \to R^n$ is locally homeomorphic, then the only obstacle to extension of the covering may be that the preimage goes to infinity. If there are sufficiently many directions (or curves) with the obstacles of this kind, then in the range space there will be a family $\tilde{\Gamma}$ of curves with positive modulus. The preimages of the curves in $\tilde{\Gamma}$ will go to infinity; if the capacity of the domain of the mapping equals zero at infinity (this is true, e.g., for $R^n$), then the modulus of the family $\Gamma$ corresponding to $\tilde{\Gamma}$ equals zero. But if $M(\Gamma) = 0$, and, moreover, the mapping changes the modulus boundedly, then $M(\tilde{\Gamma}) = M(f(\Gamma)) = 0$, and hence there would not be too many obstacles to continuation.

Although we have not given the technical details of the argument, the above sketch provides us with an idea of the interaction between quasiconformality and conformal invariants.

We now consider some generalizations of the global homeomorphism theorem or global invertibility theorem of quasiconformal mappings.

# 5 Asymptotics of admissible growth of coefficient of quasiconformality

Let $f : R^n \to R^n$ $(n \geq 3)$ be an arbitrary locally homeomorphic mapping. We denote by $k(r)$ its coefficient of quasiconformality in the ball $B^n(r) = \{x \in R^n | \; |x| = r\}$.
The following theorem is valid [Zo4].

**Theorem**

i) *If $\int^\infty \frac{dr}{rk(r)}$ diverges then the conclusion of the theorem on global homeomorphism is valid.*

ii) *In the sense of admissible order of growth of $k(r)$ the assertion i) is sharp, i.e., for any nond ecreasing function $\phi(r) > 1$ for which $\int^\infty \frac{dr}{r\phi(r)}$ converges, one can also construct a homeomorphic mapping $f$ of the space $R^n$ to a unit ball such that $k(r) < \phi(r)$ for all sufficiently large values $r$.*

The mapping mentioned in ii) can be written explicitly:

$$f(x) = \begin{cases} \frac{1}{2}x & \text{when } |x| lec, \\ \frac{1}{2}\left(1 + \left(\int_c^\infty \frac{dt}{t\phi(t)}\right)^{-1} \int_c^{|x|} \frac{dt}{t\phi(t)}\right)\frac{x}{|x|} & \text{when } |x| > c, \end{cases}$$

where $c$ is a constant chosen below. It is easy to calculate that for the mapping $f$

$$k(r) = \left(\int_c^\infty \frac{dt}{t\phi(t)} + \int_c^r \frac{dt}{t\phi(t)}\right)\phi(r)$$

for any sufficiently large values of $r$.
If we now choose the constant $c$ to obey the condition

$$\int_c \frac{dt}{t\phi(t)} < \frac{1}{2},$$

we get the desired mapping.
Part i) of the theorem provides the condition $M(\Gamma_\infty) = 0 \Rightarrow M(f(\Gamma_\infty)) = 0$ discussed in section 4, the condition of modulus transformation which, as was already mentioned above, guarantees the invertibility of $f$.
The indicated condition on modulus transformation can be provided even under weaker integral conditions [P1] than in i) . That is why the theorem on global homeomorphism is still valid for some classes of mappings $f : R^n \to R^n$ $(n \geq 3)$ that are quasiconformal only in the mean.

# 6 Removable singularities

Since the topological structure of a locally homeomorphic mapping $f : R^n \to R^n$ is defined by its behavior in a neighborhood of infinity it is natural to study [AM], [Zo5] conditions of removability of an isolated singularity of the mapping.

**Theorem** *Let $f : \dot{U} \to R^n$ be a locall y homeomorphic mapping defined in a punctured neighborhood $\dot{U} = \{x \in R^n | 0 < |x| < r_0\}$ of the point $0 \in R^n$. Let $k(r)$ be the coefficient of quasiconformality of $f$ in a domain $\{x \in R^n | 0 < r < |x| < r_0\}$. Then the following assertions are valid*

i) *If $\int_0 \frac{dr}{rk(r)} = \infty$ and $n \geq 3$, then the mapping $f$ is homeomorphic in some punctured neighborhood $\dot{V} \subset \dot{U}$ of the point $x = 0$, and it can be extended to a homeomorphism of some complet e neighborhood of this point.*

ii) *In the sense of asymptotics of admissible growth of $k(r)$, assertion i) is sharp.*

A detailed explanation of the meaning of the analogous assertion ii) was presented in the previous theorem.

This theorem on the removal of an isolated singularity is valid also for the mappings $f : \dot{U} \to R^n$ that are quasiconformal in the mean [P2].

The following question, however, is still open: *whether one can replace a point by a one-dimensional segment in the theorem on isolated (removable) singularities* for a local homeomorphism with bounded distortion, e.g., in the 3-dimensional case.

On the other hand, one can ask about the role of $R^n$ in the triple $f : \dot{U} \to R^n$. More precisely, *whether the theorem is valid for the mappings $f : \dot{U} \to M$ to an arbitrary simply-connected Riemannian manifold $M$?* Note that after factorization of $R^n \setminus \{0\}$ with respect to the relation $x \sim 2x$, we get a conformal covering of the closed manifold $S^{n-1} \times S^1$. Hence, the hypothesis of simply-connectedness is needed.

# 7 Injectivity radius

Martio, Rickman and Väisälä [MRV] obtained the following nice development of the global homeomorphism theorem, generalizing at the same time to the quasiconformal case the result due to John [J2] on quasi-isometries.

Let $Q'_k$ be a class of locally homeomorphic mappings $f : B^n \to R^n$ of the unit ball into $R^n$ with coefficient of quasiconformality $k_f < k$. Let $r_f$ be the radius of the largest ball concentric with $B^n$, where $f$ is injective.

**Theorem** (on injectivity radius) *If $n \geq 3$, then*

$$r(n, k) := \inf_{f \in Q'_k} r_f > 0. \tag{3}$$

Hence any such mapping is injective in a ball of radius $r(n, k)$, which is called the *injectivity radius*. The injectivity radius depends only on the coefficient of quasiconformality of the mapping and on the dimension of the space.

The injectivity radius obviously changes in proportion to the radius of the ball under such a transformation. Hence, replacing $B^n$ in this theorem by the whole space $R^n$ we again obtain the global homeomorphism theorem.

# 8  Quasiconformal mappings of Riemannian manifolds

All complete results discussed above in connection with the global homeomorphism theorem for quasiconformal mappings were related to the mappings defined in subdomains of $R^n$ and with range in $R^n$. Novikov pointed out an alternate direction of investigation. He stated the following question (see [Zo3]): to what extent the properties of the space affect the results. More precisely: what are the manifolds such that locally homeomorphic quasiconformal mappings of them are automatically global injections?

Gromov proposed [Gro] the following geometric version of the global homeomorphism theorem, which, in particular, can be regarded to some extent as an answer to Novikov's question.

**Theorem (on global injectivity for manifolds )** *If $f : M \to N$ is a locally homeomorphic quasiconformal mapping of a complete Riemannian manifold $M$ of finite volume into a simply-connected Riemannian manifold $N$ of dimension $n \geq 3$, then $f$ is injective and $N \setminus f(M)$ is of Hausdorff dimension zero.*

Since a complete proof of Gromov's nice assertion is not published we make some more detailed comments on it [Zo7].

If $M$ is a compact (closed) manifold, then one can conclude, just from the theorem on covering homotopy and using no assumptions of quasiconformality, that $f$ is a homeomorphism . Hence, the only nontrivial case is the case when $M$ is a complete, but noncompact, manifold. In this case it makes sense to speak about its capacity at infinity. The lengths of the curves going to infinity are infinite ($M$ is complete). Since any positive constant is an admissible function for the family of curves on the manifold going to infinity, and since the volume of $M$ is finite, one can conclude that the capacity of $M$ at infinity equals zero (see section 4).

By using the modulus properties [F], one is able to prove a converse assertion: if the capacity of a complete Riemannian manifold equals zero at infinity, then in the class of metrics conformally equivalent to the initial metric of the Riemannian manifold under consideration there exists a metric for which the manifold is still complete but has a finite volume. For example, $R^n$ is a complete Riemannian manifold of finite volume with respect to the metric $\lambda(x)|dx|$, where $\lambda(x) \sim \frac{1}{|x|\ln|x|}$ as $|x| \to \infty$.

A conformally invariant interpretation of the geometric conditions of completeness and finiteness of volume of the manifold $M$ permits us to conclude that the set $N \setminus f(M)$ omitted under the mapping $f : M \to N$ has zero capacity, and that the manifolds $f(M)$ and $N$ also have zero capacity at infinity.

In neighborhoods of infinity on the manifolds under consideration, one can introduce a (quasi)conformal (quasi)invariant metric in such a way that $f$ will become a quasi-isometry admitting the natural conformally–invariant continuation up to a homeomorphism $\bar{f} : \bar{M} \to \bar{N} = f(\bar{M})$ of the completions of $M$ and $N$

with respect to the corresponding metrics on $M$ and $N$.

In this interpretation Gromov's theorem includes the original variant of the global homeomorphism theorem, where $M = N = R^n$ and $\bar{M} = \bar{N} = S^n$.

The conformally invariant metric on a neighborhood of infinity mentioned above can be constructed on $M$, for example, in the following way (in this connection see [GoVo], [LF1], [LF2], [Vu2]). Let us fix a compact ball $K_M \subset M$. For the points $p, q \in M \setminus K_M$ we define a distance

$$d(p, q) = \inf_{\gamma_{p,q}} \text{cap}(K_M, \gamma_{p,q})$$

as the infimum of capacities of condensers $(K_M, \gamma_{p,q})$ in $M$, where $\gamma_{p,q}$ is a variable continuum in $M$ containing both $p$ and $q$.

# 9 Open problems

In addition to the problems mentioned above in section 6 we would like to present a group of closely related open questions [Zo6].

Let $f : D \to H$ be, generally speaking, a nonlinear smooth operator in the domain $D$ of a Hilbert (or even Banach) space $H$. Let the tangent linear operator $f'(x)$ have a continuous inverse at every point $x \in D$. This guarantees local invertibility of the operator $f$.

Recalling the definition of the coefficient of quasiconformality (see section 1) it is easy to understand that

$$k_{f'(x)} := \|f'(x)\| \cdot \|(f'(x))^{-1}\|,$$

and hence the notions of quasiconformality and the coefficient of quasiconformality of an operator can be regarded in their original sense.

It is natural to ask whether the above theorems:

> *on stability of conformal mappings,*
> *on coefficient of distortion,*
> *on global homeomorphism,*
> *on isolated singularity,*
> *on radius of injectivity,*

are still valid for nonlinear operators with bounded distortion (quasiconformal operators).

For example, an analog of the global homeomorphism theorem would provide us with the following principle of invertibility or

theorem of existence and uniqueness for nonlinear operators.

*If the operator $f : H \to H$ in a Hilbert space $H$ is locally invertible and has bounded distortion, then $f$ is globally invertible.*

Hence in this case the equation $f(x) = y$ is solvable in $H$; the solution is unique for any right-hand-side $y \in H$.

It can happen (as it does sometimes in operator theory) that specific properties of the space (Hilbert or arbitrary Banach space) are essential.

In connection with the problem stated it is natural to formulate the following conjectures on the classical finite-dimensional theory:

$$\sup_n \epsilon(n, k) =: E(k) < \infty, \tag{1'}$$

$$\sup_{n} \ \nu(n) \ =: \ \nu \ < \ \infty, \tag{2'}$$

$$\inf_{n>2} \ r(n,k) \ =: \ R(k) \ > \ 0 \tag{3'}$$

These three conjectures concern the dependence on dimension of the values $\epsilon(n,k)$, $\nu(n)$, $r(n,k)$ in (1), (2), and (3), respectively.

# 10   Some information

In connection with the problems discussed we call attention to the following circle of ideas.

a) When $n$ is fixed and the mappings are close to conformal (i.e., as $\epsilon \to 0+$ or $k \to 1$) the behavior of the functions $\mu(n,\epsilon)$, $\epsilon(n,k)$, $r(n,k)$ is qualitatively known. For example, one has the asymptotic estimate (2) for $\mu(n,\epsilon)$; $\epsilon(n,k) \to 1$ as $k \to 1$ and $r(n,k) \equiv 1$ when $k$ is sufficiently close to 1.

In [Vu3] there was given an explicit majorant for $\epsilon(n,k)$ for all $k > 1$ such that the majorant tends to 1 as $k$ tends to 1.

And now a few words concerning the dependence of these functions on dimension.

While studying and sharpening stability estimates in Liouville's theorem, Semenov showed [Se] that $\limsup_{\epsilon \to 0+} \mu(n,\epsilon)/\epsilon < C$, where the constant $C$ does not depend on $n$. If only passage to the limit here would be uniform with respect to $n$, we would obtain the inequality (2').

b) In [AVV1] it was shown that
$\epsilon(n,k) \leq \tilde{\epsilon}(k^{n-1})$; see also new distortion results in [AVV2], [AVV3], and [Vu3].

c) Although the behavior of $r(n,k)$ with respect to $n$ is very interesting, it has almost never been investigated. Of course it is easy to get a rough estimate from below that decreases exponentially as $n \to +\infty$.

It is easy to see that when $k$ is fixed but not too close to 1, an unrestricted growth of $n$ does not necessarily imply that $r(n,k) \to 1$.

In connection with the problem of estimating the injectivity radius, it is worth mentioning that there are relatively few theorems on local invertibility in analysis with concrete quantitative bounds for the injectivity radius or for the domain of one–to–one–ness of the local inverse mapping (cf. [NF], [NR1], [J1], [J2], [Pa]).

d) The injectivity theorems of complex analysis can be found in [Le], [G3], [Ge3], [AAE].

e) Investigations of injectivity for quasi-isometric mappings are presented in [J2], [J3], [MS], [G4], [Ge1].

f) General injectivity theorems related to stability properties of bi-Lipschitzian, quasiregular, and analytic mappings are presented in [MS] with the corresponding references. See also [I1].

g) For some stability estimates for isometries in a Banach space see [Ge2], [DG].

h) In connection with the problem of removability of one-dimensional segments (see section 6) note the following new result of Iwaniec [I2], of which I was kindly informed by S. Rickman. Developing the ideas of [DS], [IM] Iwaniec showed the existence of a function $d(n,k)$ such that the relation $H - \dim E < d(n,k)$ (where $H - \dim E$ is the Hausdorff dimension of the set $E$ in the $n$-dimensional unit ball $B^n$) implies that $E$ is a removable set for the class of $k$-quasiregular (not necessarily locally homeomorphic but bounded) mappings $f : B^n \setminus E \to B^n$. It is remarkable that $d(n, k)$ tends to infinity as $n$ grows without bound.

I wish to express my gratitude to G. D. Anderson, D.Burzan, M. Vuorinen and A. Zorich. This article could not have been written without their kind help.

## REFERENCES

[ AM ] S. Agard and A.Marden, *A removable singularity theorem for local homeomorphisms*, Indiana Math. J. **20** (1970), 455–461.

[ A1 ] L. V. Ahlfors, *Zur Theorie der Überlagerungsflächen*, Acta Math. **65** (1935), 157–194.

[ A2 ] L. V. Ahlfors, *Lectures on quasiconformal mappings*, Van Nostrand, 1966.

[ AB ] L. V. Ahlfors and A. Beurling, *Conformal invariants and function theoretic null sets*, Acta Math. **83** (1950), 101–129.

[ Al ] V. A. Aleksandrov, *On the Efimov theorem on differential conditions for homeomorphism*, (in Russian) Mat. Sb. **181** (1990), 183–188.

[ AVV1 ] G. D. Anderson, M. K. Vamanamurthy and M. Vuorinen, *Dimension-free quasiconformal distortion in n-space*, Trans. Amer. Math. Soc. **297** (1986), 687–706.

[ AVV2 ] G. D. Anderson, M. K. Vamanamurthy and M. Vuorinen, *Sharp distortion theorems for quasiconformal mappings*, Trans. Amer. Math. Soc. **305** (1988), 95–111.

[ AVV3 ] G. D. Anderson, M. K. Vamanamurthy and M. Vuorinen, *Inequalities for quasiconformal mappings in the plane and space*, Manuscript (1991), 1–43.

[ AAE ] F. G. Avkhadiev, L. A. Aksent'ev, A. M. Elizarov, *Sufficient conditions for finite-valence of analytic functions and their applications*, (in Russian) Mathematical Analysis, Itogi Nauki i Tekhniki, Acad. Nauk SSSR VINITI, Moscow **25** (1987), 3–121.

[ B1 ] P. P. Belinskiĭ (Belinsky), *General properties of quasiconformal mappings*, (in Russian)Izdat. "Nauka", Sibirsk. Otdelenie; Novosibirsk, 1974

[ B2 ] P. P. Belinskiĭ, *Stability in the Liouville theorem on space quasiconformal mappings*, (in Russian) ; in *Some problems of mathematics and mechanics*, Proceedings of the conference on occasion of the seventieth birthday of academician M. A. Lavrentiev, Academy of Sciences of the USSR, Siberian Branch, "Nauka" Publishing House; Leningrad, 1970, 88–102.

[ B3 ] P. P. Belinskiĭ, *On the range of closeness of space quasiconformal mapping to a conformal one*, (in Russian) Sibirsk. Mat. Zh. **14** (1973), 475–483.

[ BI ] B. Bojarski and T. Iwaniec, *Another approach to Liouville theorem*, Math. Nachr. **107** (1982), 253–262.

[ Ca ] P. Caraman, *n-dimensional quasiconformal mappings*, Editura Academiei Române; Bucharest, Abacus Press;Tunebridge Wells Haessner Publishing, Inc. ; Newfoundland, New Jersey, 1974.

[ C ] H. Cartan, *Sur les transformations localement topologiques*, Acta Litter **6** (1933), 85–104.

[ DG ] Ding Guanggui, *The approximation problem of almost isometric operators by isometric operators*, Acta Math. Scientia **8** (1988), 361–372.

[ DS ] S. K. Donaldson and D. P. Sullivan, *Quasiconformal 4-manifolds*, Acta Math. **163** (1989), 181–252.

[ F ] B. Fuglede, *Extremal length and functional completion*, Acta Math. **98** (1957), 171–219.

[ G1 ] F. W. Gehring, *Symmetrization of rings in space*, Trans. Amer. Math. Soc. **101** (1961), 499–519.

[ G2 ] F. W. Gehring, *Rings and quasiconformal mappings in space*, Trans. Amer. Math. Soc. **103** (1962), 353–393.

[ G3 ] F. W. Gehring, *Injectivity of local quasi-isometries*, Comm. Math. Helv. **57** (1982), 202–220.

[ G4 ] F. W. Gehring, *Topics in quasiconformal mappings*, Proc. Internat. Congr. Math. (Berkeley, California, 1986) **1, 2** (1987), 62–80; Amer. Math. Soc., Providence,Rl.

[ GV ] F. W. Gehring and J. Väisälä, *The coefficient of quasiconformality of domains in space*, Acta Math. **114** (1965), 1–70.

[ Ge1 ] J. Gevirtz, *Injectivity in Banach spaces and the Mazur—Ulam theorem on isometries*, Trans. Amer. Math. Soc. **274** (1982), 307–318.

[ Ge2 ] J. Gevirtz, *Stability of isometries on Banach spaces*, Proc. Amer. Math. Soc. **89** (1983), 633–636.

[ Ge3 ] J. Gevirtz, *On extremal functions for John constants*, J. London Math. Soc. Ser. 2 **39** (1989), 285–298.

[ Go ] V. M. Gol'dstein, *On the behavior of mappings with bounded distortion with distortion coefficient close to unity*, Siberian Math. J. **12** (1971), 1250–1258.

[ GoVo ] V. M. Gol'dstein and S. K. Vodop'janov, *Metric completion of a domain by using a conformal capacity invariant under quasiconformal mappings*, (in Russian) Dokl. Acad. Nauk SSSR **238** (1978), English transl. in Soviet Math. Dokl. **19** (1978), 158–161.

[ Gro ]  M. Gromov, *Hyperbolic manifolds, groups and actions*, Proceedings of the 1978 Stony Brook Conference Ann. of Math. Studies **97** (1981), 183–213 Princeton Univ. Press ; Princeton, N. J.,.

[ Gr1 ]  H. Grötzsch, *Über die Verzerrung bei schlichten nichtkonformen Abbildungen und über eine damit zusammenhängende Erweiterung des Picardschen Satzes*, (1928), **80** 503–507 Ber. Verh. Sächs. Acad. Wiss. Leipzig.

[ Gr2 ]  H. Grötzsch, *Über möglichst konforme Abbildungen von schlichten Bereichen*, Ber. Verh. Sächs. Akad. Wiss. Leipzig **84** (1932), 114–120.

[ Hd ]  M. Hadamard, *Sur les transformations ponctuelles*, Bull. Soc. Math. France **34** (1906), 71–84.

[ He ]  J. Hesse, *A p-extremal length and p-capacity equality*, Ark. Mat. **13** (1975), 131–144.

[ I1 ]  T. Iwaniec, *Stability property of Möbius mappings*, Proc. Amer. Math. Soc. **100** (1987), 61–69.

[ I2 ]  T. Iwaniec, *p−harmonic tensors and quasiregular mappings*, Manuscript (1991), 1–45.

[ IM ]  T. Iwaniec and G. Martin, *Quasiregular mappings in even dimensions*, Acta Math. (to appear). XIV Rolf Nevanlinna Colloquium, Helsinki, June 10–14, 1990. Abstracts, p. 26.

[ JV ]  P. Järvi and M. Vuorinen, *Self-similar Cantor sets and quasiregular mappings*, J. Reine Angew. Math. **242** (1992), 31–45.

[ J1 ]  F. John, *Rotation and strain*, Comm. Pure App. Math. **14** (1961), 391–413.

[ J2 ]  F. John, *On Quasi-Isometric Mappings, I*, Comm. Pure App. Math. **21** (1968), 77–103.

[ J3 ]  F. John, *On Quasi-Isometric Mappings, II*, Comm. Pure App. Math. **22** (1969), 41–66.

[ K ]  A. P. Kopylov, *Stability in some classes of mappings in C-norm*, (in Russian) Izdat. "Nauka", Sibirsk. Otdelenie ; Novosibirsk, 1990.

[ KM ]  P. Koskela and O. Martio, *Removability theorems for quasiregular mappings*, Ann. Acad. Sci. Fenn. Ser. A I Math. **15** (1990), 381-399.

[ Lav1 ]  M. A. Lavrentieff, *Sur une classe de representatations continues*, Mat. Sb. **42** (1935), 407–427.

[ Lav2 ]  M. A. Lavrentieff, *Sur un critère differentiel des transformations homéomorphes des domains à trois dimensions*, Dokl. Acad. Nauk SSSR **20** (1938), 241–242.

[ Lav3 ]  M. A. Lavrent'ev (Lavrentiev), *On stability in Liouvillle's theorem*, Dokl. Acad. Nauk SSSR **95** (1954), 925–926.

[ Lav4 ] M. A. Lavrent'ev, *On the theory of quasiconformal mappings of three-dimensional domains*, Sibirsk. Mat. Zh. (in Russian) **5** (1964), 596–602.

[ Lav5 ] M. A. Lavrent'ev, *On the theory of quasiconformal mappings of three-dimensional domains*, J. Analyse Math. **19** (1967), 217–225.

[ Le ] O. Lehto, *Univalent functions and Teichmüller spaces*, Graduate Texts in Math. **109** 1987; Springer-Verlag, Berlin—Heidelberg—New York.

[ LF1 ] J. Lelong-Ferrand, *Invariants conformes globaux sur les varietes riemanniennes*, J. Differential Geom. **8** (1973), 487–510.

[ LF2 ] J. Lelong-Ferrand, *Construction du métriques pour lesquelles les transformations sont lipschiziennes*, Symposia Mathematica **18** (1976), 407–420, INDAM, Academic Press; London.

[ Lo ] C. Loewner, *On the conformal capacity in space*, J. Math. Mech **8** (1959), 411–414.

[ MRV ] O. Martio, S. Rickman and J. Väisälä, *Topological and metric properties of quasiregular mappings*, Ann. Acad. Sci. Fenn. Ser. A. I Math. **488** (1971), 1–31.

[ MSa ] O. Martio and J. Sarvas, *Injectivity theorems in plane and space*, Ann. Acad. Sci. Fenn. Ser. A I Math. **4** (1979), 383–401.

[ MSr ] O. Martio and U. Srebro, *Universal radius of injectivity for locally quasiconformal mappings*, Israel J. Math. **29** (1978), 17–23.

[ N1 ] R. Näkki, *Boundary behavior of quasiconformal mappings in n-space*, Ann. Acad. Sci. Fenn. Ser. A I **484** (1970), 1–50.

[ N2 ] R. Näkki, *Prime ends and quasiconformal mappings*, J. Analyse Math. **35** (1979), 13–40.

[ NF ] F. Nevanlinna, *Über die Umkehrung differenzierbarer Abbildungen*, Ann. Acad. Sci. Fenn. Ser. A I Math. **245** (1957), 1–14.

[ NR1 ] R. Nevanlinna, *Über die Methode der sukzessiven Approximationen*, Ann. Acad. Sci. Fenn. Ser. A I Math. **291** (1960), 1–10.

[ NR2 ] R. Nevanlinna, *On differentiable mappings*, Analytic functions. Princeton Math. Series. **24** (1960), 3–9, Princeton Univ. Press;Princeton.

[ Pa ] T. Parthasarathy, *On global univalence theorems*, Lecture Notes in Mathematics **977** (1983), Springer-Verlag.

[ P1 ] M. Perovich (Perović), *On the global homeomorphism of the mean quasiconformal mappings*, (in Russian) Dokl. Acad. Nauk SSSR **230** (1976) 781–784; English transl. in Soviet Math. Dokl. (1976).

[ P2 ] M. Perovich, *Isolated singularity of the mapping quasiconformal in the mean*, Proceedings of the III Romanian-Finnish Seminar (1976) Lecture Notes in Math. **743** (1979), 212–214, Springer—Verlag.

[ Re1 ]  Yu. G. Reshetnyak, *Liouville's theorem on conformal mappings under minimal regularity assumptions*, (in Russian) Sibirsk. Mat. Zh. **8** (1967), 835–840.

[ Re2 ]  Yu. G. Reshetnyak, *Stability theorems in geometry and analysis*, (in Russian) Izdat. "Nauka", Sibirsk. Otdelenie ; Novosibirsk, 1982.

[ Re3 ]  Yu. G. Reshetnyak, *Space Mappings with Bounded Distortion*, Translations of mathematical Monographs **73** (1989), Amer. Math. Soc.; Providence, R.I..

[ Ri1 ]  S. Rickman, *On the number of omitted values of entire quasiregular mappings*, J. Analyse Math. **37** (1980), 100–117.

[ Ri2 ]  S. Rickman, *The analogue of Picard's theorem for quasiregular mappings in dimension three*, Acta Math. **154** (1985), 195–242.

[ Ri3 ]  S. Rickman, *Nonremovable Cantor sets for bounded quasiregular mappings*, Institut Mittag–Leffler Report **42** (1989/90).

[ Se ]  V. I. Semenov, *An integral representation of the trace of the sphere of a class of vector fields and uniform stability estimates of quasiconformal mappings of the ball*, Mat. Sb. **133** (1987), 238–253.

[ Sh ]  B. V. Shabat (Šabat), *The modulus method in space*, (in Russian) Dokl. Acad. Nauk SSSR **130** (1960) 1210–1213; English transl. in Soviet Math. Dokl. **1** (1960), 165–168.

[ Sy ]  A. V. Sychev, *Moduli and n-dimensional quasiconformal mappings*, (in Russian) Izdat. "Nauka", Sibirsk. Otdelenie ; Novosibirsk, 1983.

[ T1 ]  O. Teichmüller, *Untersuchungen über konforme und quasikonforme Abbildung*, Deutsche Math. **3** (1938), 621–678.

[ T2 ]  O. Teichmüller, *Extremal quasikonforme Abbildungen und quadratische Differentiale*, Abh. Preuse. Akad. Wiss. Mat.-Nat. Kl. **22** (1940), 1–197.

[ V1 ]  J. Väisälä, *On quasiconformal mappings in space*, Ann. Acad. Sci. Fenn. Ser. A I Math. **298** (1961), 1–36.

[ V2 ]  J. Väisälä, *On quasiconformal mappings of a ball*, Ann. Acad. Sci. Fenn. Ser. A I Math. **304** (1961), 1–17.

[ V3 ]  J. Väisälä, *Lectures on n-dimensional quasiconformal mappings*, Lecture Notes in Mathematics **229** (1971), Springer-Verlag.

[ Vu1 ]  M. Vuorinen, *On the boundary behavior of locally K-quasiconformal mappings in space*, Ann. Acad. Sci. Fenn. Ser. A I Math. **5** (1980), 79–85.

[ Vu2 ]  M. Vuorinen *Conformal Geometry and Quasiregular Mappings*, Lecture Notes in Mathematics **1319** (1988), Springer-Verlag.

[ Vu3 ]  M. Vuorinen, *Quadruples and spatial quasiconformal mappings*, Math. Z. **205** (1990), 617–628.

[ Zi ]  W. P. Ziemer, *Extremal length as a capacity*, Mich. Math. J. **17** (1970), 117–128.

[ Zo1 ]  V. A. Zorich (Zorić), *Boundary correspondence under Q-quasiconformal mappings of a ball*, (in Russian) Dokl. Acad. Nauk SSSR **145** (1962), 1209–1212 ; English transl. in Soviet Math. Dokl. **3** (1962), 1183–1186.

[ Zo2 ]  V. A. Zorich, *Homeomorphity of space quasiconformal mappings*, (in Russian) Dokl. Acad. Nauk SSSR **176** (1967), 31–34 ; English transl. in Soviet Math. Dokl. **8** (1967), 1039–1042.

[ Zo3 ]  V. A. Zorich, *The theorem of M. A. Lavrent'ev on quasiconformal space mappings*, (in Russian) Mat. Sb. **74** (1967), 417–433 ; English transl. in Math. USSR — Sbornik **3** (1967), 389–403.

[ Zo4 ]  V. A. Zorich, *Admissible order of growth of the quasiconformal characteristic in Lavrent'ev's theorem*, (in Russian) Dokl. Acad. Nauk SSSR **181** (1968) ; English transl. in Soviet Math. Dokl. **9** (1968), 866–869.

[ Zo5 ]  V. A. Zorich, *An isolated singularity of mappings with bounded distortion*, (in Russian) Mat. Sb. **81** (1970), 634–638 ; English transl. in Math. USSR — Sbornik **10** (1970), 581–583.

[ Zo6 ]  V. A. Zorich, *Bounded dependence on dimension of some important quantities of the theory of space quasiconformal mappings*, (in Russian) Mat. Vesnik **40** (1988), 371–376.

[ Zo7 ]  V. A. Zorich, *On Gromov's geometric version of the global homeomorphism theorem for quasiconformal mappings*, XIV Rolf Nevanlinna Colloquium, Helsinki, June 10-14, 1990. Abstracts, p. 36. (Complete text in preparation).